Advances in Anatomy Embryology and Cell Biology

Vol. 163

AF842273

Springer

Berlin
Heidelberg
New York
Barcelona
Hong Kong
London
Milan
Paris
Tokyo

J. N. Maina

Fundamental Structural Aspects and Features in the Bioengineering of the Gas Exchangers: Comparative Perspectives

With 142 Figures

Springer

J. N. Maina
Department of Anatomical Sciences,
The University of the Witwatersrand,
7 York Road, Park Town 2193,
Johannesburg, South Africa

ISSN 0301-5556
ISBN 3-540-42951-4 Springer-Verlag Berlin Heidelberg NewYork

Library of Congress-Cataloging-in-Publication-Data
Maina, John N.: Fundamental structural aspects and features in the bioengineering of the gas exchangers: comparative perspectives / J. N. Maina. – Berlin; Heidelberg; New York; Barcelona; Hong Kong; London; Milan; Paris; Tokyo: Springer, 2002
(Advances in anatomy, embryology, and cell biology; Vol. 163)
ISBN 3-540-42951-4

Springer-Verlag a member of BertelsmannSpringer
Science + Business Media GmbH

http://www.springer.de

Production: PRO EDIT GmbH, 69126 Heidelberg, Germany
Printed on acid-free paper – SPIN: 10848515 27/3130wg - 5 4 3 2 1 0

Though not very often stated, explicitly or implicitly, the aim of biological science is to formulate the laws that govern cells, organisms, and societies of organisms. Through multifaceted approaches, biologists subscribe to the notion that their results will ultimately be explicable at cellular and molecular levels.

Gordon (1988)

Preface

> The history of biology is replete with examples of how comparative biology helped clarify the meaning of structure and function in complex animals. Indeed, without the comparative approach to biology, the birth of physiology would have been delayed.
>
> Fishman (1979)

Comparative morphologists are challenged to discern the changes that have occurred in evolution and development of the forms and states of organisms as well as to explain the factors that compelled them (e.g. Dullemeijer 1974). The main objective of this contribution is to present what I deem to be some of the fundamental structural aspects in the design of respiratory organs while debating and speculating on when, how and why these states were founded. My main thesis is that the modern gas exchangers are products of protracted processes that have entailed adaptation to specific environments and lifestyles. Only those feasible designs that have proven adequately competent in meeting demands for molecular oxygen have been preserved. Unfortunately, August Krogh's (Krogh 1941) and Pierre Dejours' (Dejours 1975) seminal works on the comparative physiology of the respiratory organs have not been paralleled by equally extensive and detailed morphological work. Our approach has been to look into the limiting functional properties as regards the respiratory capacities of gas exchangers while finding out the specific structural adaptations that have evolved to meet the metabolic needs or to look into form and to discern how it limits function. This has allowed a deduction of structure–function correlation. Since they offer novel solutions to respiratory challenges, 'extreme animals' have received particular attention. Notwithstanding the uncertainties that accompany extrapolative studies like ours, we believe that we have developed satisfactory conceptual models on which the multiplicity of the designs of the gas exchangers can be rationalized. Animals endowed with different infrastructural resources have, through extraordinary bioengineering feats, accomplished remarkably congruent designs. For biologists, there is always some unique kind of satisfaction when a hitherto-unknown aspect of biology fits into or even deviates from a theoretical paradigm, a morphological prediction, a physiological process or a developmental principle. It should, however, now be appreciated that it is no longer necessary to formulate a scientific question to suit a particular animal. A transgenic animal can, for example, be appropriately designed to best answer a particular question.

Morphology has been pivotal in understanding the biology of living organisms. Exploring and explicating relationships between biological processes and structures rationalizes and demystifies the scientific discipline. Since soft tissues are very seldomly fossilized and, if fossilization does occur, they are flattened by compression, making their recognition and interpretation difficult, the historical events and processes that have shaped the design of the gas exchangers can only be connected indirectly. This entails deduction based on analogous and homologous structures and, inevitably, perceptive teleological reasoning. A sound scientific basis for debating the 'paleomorphology' of gas exchangers is possible. Although most of the views expressed in this book are presently most probably correct, I am under no illusion that I have espoused certain ideas that sooner or later will be shown to be absolete; they may be totally or partly revised or even rejected. This is because, realistically, the current data are insufficient to secure absolute sustainability of certain observations. A lot of work still remains to be done to consolidate the available knowledge on the various aspects of the evolution and design of gas exchangers. Important relevant areas such as comparative biochemistry and comparative molecular evolution of the respiratory structures are unfortunately nonexistent or still in their infancy. I, nevertheless, hope that the views and concepts offered will provoke debate or, better still, prompt further studies. In science, one can only argue upon the available facts. As further investigations are performed and more innovative techniques developed, new data may be adduced. If a different understanding emerges, the 'story' must change. As Gil-Vernet (1968) pointed out, 'science has no end'. It is essentially a process of 'passing the baton' from one individual to another. In a theatre of continuous inquiry and presentation of facts, debate and exchange of views and opinions, it is gratifying and even inspiring to note that science is a common enterprise that thrills all humankind.

Johannesburg, December 2001 J. N. Maina

Contents

Acknowledgements

I have been extremely fortunate to work with most helpful, delightful and enthusiastic colleagues. All cannot be mentioned here. The foremost include Prof. A.S. King (Emeritus Prof.), University of Liverpool; Dr. M.A. Abdalla, King Saud University; and Prof. G.M.O. Maloiy, University of Nairobi. Research grants and fellowships that have greatly advanced my research work have been kindly given by the British Council, The Royal Society of the United Kingdom, The Fulbright Program, The Leverhume Trust, Canadian International Development Agency (CIDA), National Scientific and Engineering Research Council of Canada (NSERC), The University of Nairobi and The University of the Witwatersrand. I dedicate this work to my family: Wanjuku, Ndegwa, Wanjiru and Kireru.

1 Introduction: Evolution of the Respiratory Processes

Organisms are not exempt from the rules of physics and chemistry. They are constrained in their evolution of gas exchangers, development of respiratory strategies and exploitation of environments by those characteristics peculiar to oxygen, carbon dioxide, air, water, and the other constituents of gas exchange.

Prange (1996)

In comparative functional morphology, the primary objective is to seek to understand the essence of the founding of the modern animal body forms and sizes by explicating the integrative themes and principles that spatially (horizontally) diffuse across taxa and temporarily (vertically) filter through the evolutionary continuum. Such inclusive data should offer robust answers regarding the basis of organismal and organic designs while underscoring the constraints faced and the trade-offs and compromises transacted towards the development of the definitive morphologies. In this account, a critical synthesis of the constructs of the gas exchangers is given while arguing on the permutations that occurred and the factors that prescribed the inauguration of certain designs in preference to others. The infrastructural resources requisitioned and committed to the construction of the different gas exchangers are comparatively highlighted.

Since Hippocrates (460–377) first proposed that the main purpose of breathing is to cool the heart and Antoine Lavoisier (1743–1794) and Joseph Priestley (1733–1804) figured out that animals breathe air to acquire O_2, respiratory biology has advanced tremendously. Breathing has become such an integral physiological process for life that it is generally taken for granted. As animals go about their daily activities, very

[1] In this treatise, the word 'morphology' is not used simply as a synonym of anatomy, i.e. description of structure for its own sake (as the word is unfortunately commonly used today), but rather as understood by D'Arcy Thompson (Thompson 1959) that 'morphology is not a study of material things and forms of material things, but has its dynamical aspect, under which we deal with the interpretation, in terms of force, of the operations of Energy'. That approach of seeking to understand the functional basis of structure reflects the true etymology of the word 'morphology'.

[2] The term 'design', which is borrowed from engineering, is used in this monograph to mean 'creative natural arrangement of parts [= components in a device (gas exchanger) for a particular purpose (respiration)]'. Vogel (1988) defined 'biological design' as 'functionally competent arrangement of parts resulting from natural selection'.

[3] The term 'gas exchanger' is used here to encompass all structures that have either primarily developed for, or have secondarily acquired, a respiratory role irrespective of the animal species in which they occur, their location, structural complexity and the respiratory medium utilized.

little attention or conscious effort is given to the air that they breathe, and even how they breathe it, until or unless something goes wrong with its availability, acquisition or composition. Instantly, their very existence is severely threatened. Although animals will generally live for weeks without food and for days without water, O_2 is continually needed for energy production. Only the most primitive forms of life (e.g. the prokaryotic anaerobes) can survive without O_2. Since its emergence and subsequent accretion, O_2 has been the most singularly pervasive molecular factor that has set the adaptive radiation, the speciation, the ecological distribution and the geographical patterning of animal life. In their classic works, August Krogh (Krogh 1941) and Max Kleiber (Kleiber 1961) termed the acute need for molecular O_2 for life 'the call for oxygen' and 'the fire of life', respectively. More recently, Laitman et al. (1996) declared that 'the acquisition and processing of O_2 and its by-products is the primary mission of any air-breathing vertebrate'.

Constituting about one-quarter of all the atoms in the living matter, O_2 has played a fundamental role as a building block of virtually all the vital organic molecules. In the nearly 4 billion years (1 billion years=10^9 years) that life has subsisted on Earth, the development, the structure and the function of the respiratory organs are inscribed in the history of molecular O_2, i.e. its absence and availability in the biosphere. The levels have shifted dramatically from neutral to reducing and from anoxia, hypoxia to hyperoxia (not necessarily in that order) relative to the modern state (e.g. Graham et al. 1995). Paradoxically, the biochemistry of molecular O_2 appears to be irreconcilable with life. Because of its highly oxidizing property, life could only have begun in an anoxic environment. Once formed, however, despite the great toxicity of its reactive species on carbon-based life, life became intricately dependent on molecular O_2 for energy (ATP) production. In the metabolic pathways, O_2 mops up protons (H^+) to form water, an innocuous and essential chemical factor for life. Moreover, molecular O_2 is an important resource for body growth and development. Unlike metabolic substrates such as carbohydrates and fats that can be stored in large quantities in the body and used as needed, O_2 has to be continually outsourced. In the human being, about 12,000 l of air passes through the lung everyday (e.g. Burri 1985). At any one moment, however, a 70-kg person only has 1.55 l of O_2 in the body. Of that, 370 cm^3 is found in the alveoli, about 280 cm^3 in the arterial blood, about 600 cm^3 in the capillary and venous blood, 60 cm^3 is dissolved in the body tissues and 240 cm^3 is chemically bound to the muscle myoglobin. At rest, the total amount of O_2 dissolved in the tissues (~0.8 $cm^3.kg^{-1}$) can support life for a maximum period of about 6 min and for a far shorter time during exercise (e.g. Snyder 1983).

Three basic designs of the gas exchangers, namely, simple (plain) surface (e.g. cell membrane and skin), evaginated (e.g. gills) and invaginated (e.g. lungs) schemes and various combinations, have evolved in the Animal Kingdom (Fig. 1): mixed conditions generally occur in the transitional (= bimodal = amphibious) breathers. Considering that the diversity of the extant animal life may range from 50 to 100 million species, the morphological similitude in the design of gas exchangers is remarkable. The exceptional convergence unequivocally underscores the import of respiration for life. Although animals occupy different habitats as regards states and conditions such as temperature and food availability, there are only two naturally occurring respirable fluid media from which O_2 can be procured – water and air. Animals had to adapt to one or the other and in rarer cases to both. From the perspectives of the morphoarchitectonics and the engineering of the gas exchangers, the profound analogy attests to the fact that only particular designs and constructions are feasible and appropriate for respiration. The structural and functional congruity between the avian and insectan respiratory systems {both taxa have air sacs and a coherent system of air conduits [trachea and tracheoles in insects (Figs. 67–77, 102, 103) and primary, secondary and tertiary bronchi in birds (Figs. 5, 6, 17, 63, 100)] that are ventilated continuously and unidirectionally (e.g. Weis-Fogh 1964a; Maina 1989a)}, in extremely successful taxa that are separated by some 250 million years of evolutionary time [insects evolved about 400 million years ago (mya) (e.g. Callahan 1972) and birds 150 mya (e.g. de Beer 1954)], present a masterpiece of morphological and functional convergence and analogy. Although it could be argued that the high metabolic demands for flight compelled development of exceptionally efficient and congruous respiratory organs in the two taxa, interestingly, when bats evolved some 100 million years after birds [the oldest known bat is apparently the 50-million-year-old *Icaronycteris index* of the Eocene (e.g. Wimsatt, 1970)], they retained but highly refined the basic structure of the mammalian lung (e.g. Maina et al. 1982; Maina and King 1984; Maina 1985, 1986; Maina et al. 1991). In addition, bats greatly improved the cardiovascular system, particularly regarding the O_2 uptake of blood (e.g. Jürgens et al. 1981). From the above structural and functional manifestations, it can be concluded that there is no absolute prescriptive design of the respiratory organ(s) for flight or, for that matter, any other lifestyle. By adopting a broad-based adaptive stratagem where a range of organs and organ-systems were integrated in the respiratory process, a drastic overhaul of the mammalian lung for flight does not appear to have been obligatory in bats. That option may have been less risky and less costly compared with the narrower approach of highly fine-tuning the respiratory organs adopted by insects and birds. In insects, the cardiovascular and respiratory systems were uncoupled, the former being relegated to an insignificant role in the respiratory process. Wagner (1998) called the biological structures that have remained unchanged for a long time under different selective pressures the 'frozen core or the Bauplan'. Such immutable states must be of great importance for survival. They should be of special interest to functional morphologists and physiologists. The highly defended structures must depict those factors that obliged preference for and requisitioning of certain respiratory morphologies in preference for others. Hochachka et al. (1999) contended that once established, it is more difficult to keep complex physiological systems the same through a long evolutionary time than to allow them to change in response to selective pressures. Since physiological processes are executed by discrete morphological structures, it is reasonable to speculate that it is equally more costly to conserve structural designs than to allow them to evolve. This must occur only if such features are critical for life. We have endeavoured to identify the invariant design features that have been fundamental in the construction of the gas exchangers.

In its most basic form, a mechanistic model of a gas exchanger is one in which respiratory media are separated by a physical (tissue) barrier across which flux of respiratory gases, i.e. O_2 and CO_2, occurs under a pressure gradient. Such a design is presented in its entirety in the unicells (Fig. 1A), where the intracellular and extracellular compartments are separated by a cell membrane. When extrapolated to complex gas exchangers, although inherently oversimplistic, the single cell model offers a useful basis for analysing and comparing the structural endowments of the evolved gas exchangers. The diffusing capacity (i.e. the conductance) of O_2 by a gas exchanger (Do_2) is defined as the transfer of O_2 from the external respiratory medium (water and/or air) to the internal medium (blood) per unit time per unit partial pressure gradient (e.g. Roughton and Forster 1957). According to Fick's law, the conductance of a gas exchanger is determined by factors such as the respiratory surface area (Sect. 10), the prevailing partial pressure gradient, the thickness of the water/air–blood (tissue) barrier (Sect. 11; Fig. 2), the volume of the capillary blood (Sect. 12) and the permeability of the tissue compartments across which O_2 diffuses. In most complex gas exchangers, a construction that involves thin parallel epithelial cells adjoined by cellular and/or connective tissue elements, a sheet flow design, occurs (e.g. Maina 1998a, 2000a,b) (Sect. 15). The design impels the blood to flow through a maze of fine channels, spreading into an extremely thin film (e.g. Maina, 2000b). The functional consequence of the construction of a gas exchanger can be elucidated by modelling, i.e. by mathematically integrating the structural parameters that determine the gas exchange capacity (e.g. Maina et al. 1989a). When integrated with permeation constants, this converts a steady-state model to a dynamical one that expresses the respiratory capacity of a gas exchanger for O_2 per unit time per unit partial pressure gradient. 'Agitating' the model gives insights into the effects that individual and aggregate components have on the performance of a gas exchanger.

Nature has been particularly inventive in the fabrication of optimal gas exchangers. Factors such as the respiratory medium utilized, habitat occupied, phylogenetic level of development and lifestyle pursued have together prescribed the ultimate designs and constructions. The economics of infrastructural commitment, operational and maintenance cost versus application and usage (i.e. the returns), were devised and honed by nature a long time ago. Above the incipient nondescript cell membrane of the microorganisms (Fig. 1A), the respiratory organs/structures prescriptively display all or some of the following morphological features:

a Invagination (infolding=cavitation) or evagination (outfolding) from the body surface
b. Compartmentalization and/or stratification
c. Vascularization
d. Thin partitioning between the internal and external compartments that hold the gas exchange media

Reconstruction of the evolutionary processes that the gas exchangers have followed is necessary for meaningful rationalization and understanding of the basis of the similarities and differences in the designs of modern respiratory organs. Because of the lack of fossilized materials from soft structures such as the gas exchangers, the task of extrapolating is formidable and somewhat uncertain. Notwithstanding the impediments the changes that have resulted and the sequence of their occurrence can be pieced together cautiously by studying the respiratory organs of the following organisms/animals:

a. Those that occupy unique habitats as regards availability of respiratory gases
b. Those that exhibit singular behavioural activities and metabolic capacities
c. Those that display transitional respiration, i.e. use different gas exchangers to procure O_2 from air and water
d. Those at different phylogenetic levels of development
e. Those that manifest unique developmental transformations in their life histories and shifts in respiratory pathways and mechanisms

In both aquatic and terrestrial animal life, the growth and progression of the morphologies of the gas exchangers have occurred in direct response to the prevailing demands for molecular O_2 (e.g. Maina 1994, 1998a). Except in the small tracheates, where O_2 is delivered directly to the tissue cells by a contiguous system of air conduits, the trachea and tracheoles, in all complex animal life (Metazoa) in which conspicuous respiratory site(s) appear, O_2 is delivered from the external respiratory medium to the mitochondria through an intricate pathway that comprises various tissue compartments. A fluid respiratory medium (water/air) in which O_2 is dissolved is convectively transported to the respiratory site from where it diffuses across a tissue barrier (Fig. 2). In most animals, O_2 is reversibly chemically bound to a carrier (metal based) pigment and transported by blood. At the blood capillary level, it diffuses into the surrounding tissue spaces, across the cell membranes and ultimately into the mitochondria (the terminal O_2 sink), where it is engaged in the production of energy (ATP). During its flow, the partial pressure of oxygen (Po_2) drops from about 20 kPa to almost zero in the vicinity of mitochondria of the distant tissue cells (e.g. Graiger et al. 1995). Between the capillaries of the heart muscle and the mitochondria, the Po_2 drops by 2.7 kPa (e.g. Tamura et al. 1989) and between the cytosol and the mitochondria by less than 0.03 kPa (e.g. Clark et al. 1987.

In all gas exchangers, the movement of O_2 across tissue barriers occurs entirely by diffusion. As a factor that must be continually requisitioned for life, active uptake of O_2, as was thought to occur in the vertebrate lungs at the turn of the twentieth century (see Haldane 1922), would conceivably obligate enormous energy expenditure, rendering the process uneconomical. Moreover, considering the large amount of O_2 that is needed, particularly under stressful conditions such as exercise, a design obligating active acquisition of O_2 would perhaps be untenable within the practicable designs of the evolved gas exchangers. Flying insects achieve the highest mass-specific rates of aerobic metabolism in the animal kingdom (e.g. Suarez 2000). Only a highly efficient respiratory system could support such extreme aerobic capacities. The insectan tracheal system delivers O_2 from the atmosphere directly to the tissue cells (e.g. Weis-Fogh 1964a,b). In general, in the large insect species and in certain small species during exercise, convective movement of air effected by abdominal and to a lesser extent thoracic pumping assists in the movement of O_2 down the tracheal system: The supply of O_2 by the tracheal system can be ten times higher per gram of body tissue than that which the vertebrate blood capillary system can supply. Development of auxiliary respiratory organs on decalcified areas on the meral segments of the walking legs of the porcelain crabs (e.g. Stillman 2000) has allowed the taxon to attain larger body sizes and higher levels of metabolism.

Important morphological and physiological transformations of gas exchangers occurred at certain points in the evolutionary history of animal life. The changes conceivably took place when the metabolic needs of organisms exceeded those that could be sufficiently serviced by the default gas exchangers or when the biosphere changed drastically, especially with regard to levels and availability of O_2. Gas exchangers were deconstructed and reconstructed, and new ones were founded. Chronologically, the consequential events were:

a. The shift from inefficient fermentative metabolism of the primordial prokaryotes to the more efficient aerobiotic metabolism of the eukaryotes

b. The development of multicellular (complex) organisms through accretion of cells, a process that necessitated formation of definite respiratory sites and efficient convection and perfusion mechanisms to promote transport of O_2 to 'distant' sites
c. The elaboration of a closed circulatory system from an open system, a scheme that enhanced the flow and the return of haemolymph/blood to the heart and hence its transit through the gas exchangers
d. The development of various types of metal-based carrier pigments in blood and in fixed tissues (e.g., myoglobin in muscle tissue) that promoted uptake, storage and transfer of O_2
e. The development of invaginated respiratory organs (lungs) from evaginated ones (gills), a transformation accompanied by drastic changes in the modalities of ventilation and perfusion
f. The transition from use of water to air, fluid respiratory media of different physical properties
g. The development of a double circulation, a state that was accompanied by alignment of the heart in series with the lungs
h. The conversion from buccal force pumping to suctional breathing, a more effective and less energetically expensive means of convective movement of air to the respiratory site (lung)
i. The establishment of endothermic homeothermy from ectothermic heterothermy, an energetically costly state that necessitated more efficient gas exchangers
j. The development of highly energetic lifestyles (e.g. flight) that called for highly efficient supporting gas exchangers
k. Adaptation to unique habitats, e.g. hypoxic or anoxic ones

Owing to the fact that the major developmental changes of the gas exchangers have occurred through 'quantum leaps', when analysing the evolution of respiratory organs, progressive (i.e. gradual) morphological change from one form to another should not be expected. For a state prescribed by many factors (i.e. a highly permutative process), the differences or similarities should bear little correspondence, if any, to the phylogenetic histories of animal taxa. For example, although mammals diverged much earlier from the reptilian lineage than birds (e.g. Pough et al. 1989), the shape of the alveoli of the mammalian lung (Figs. 78, 79) more closely resembles that of the faveoli of the reptilian lung (Fig. 83) than the air capillaries of the avian lung resemble faveoli (Fig. 65). In fact, regarding shape and size, among the evolved gas exchangers, the air capillaries (Figs. 57–62, 65, 113) are unique to the avian lung. It could be argued that the avian lung is so structurally unique that it could not possibly have evolved from a reptilian one. The extreme selective pressure for attainment of flight, a highly energetically demanding form of locomotion (e.g. Thomas 1987), partly compelled the novel design. Unlike the brain, the progressive structural complexity and functional capacity of which reasonably develops up the evolutionary hierarchy, the design of the gas exchangers appears to have responded only to needs and circumstances and very little to phylogenetic advancement. For example, among the evolved gas exchangers, the bronchoalveolar lung of the human being is far from being the most efficient (Sect. 9; Fig. 3).

2 Metabolic Demands and Design of Gas Exchangers

Physicists inhabit a universe of abstract concepts such as systems, entropy, sources, and sinks. The biologist's world is populated by singular and wriggly objects such as fungi and butterflies, genes and enzymes, and it is not always obvious how the former impinges upon the latter.

Harold (1986)

Metabolic rate determines vital aspects of life such as population dynamics, behavioural ecology and reproductive success in animals. It expresses the integral speed at which energy is mobilized, transformed and utilized for biological processes and activities. The rate and efficiency of procuring and utilizing O_2 for metabolic purposes bespeaks the speed at which an animal uses its resources to meet the demands placed on it by the environment and the lifestyle that it pursues.

Body size and shape are not mere vagaries of nature. Rather, physical laws acting upon the phenotype prescribe them. In modern ecosystems, chance of extinction correlates directly with body size (e.g. Carroll 1988). Predator avoidance corresponds with the energy that an animal can muster to effect evasive action (e.g. Peters 1983). The capacity to procure, transport and utilize large amounts of O_2 was a major factor in the evolution and adaptive radiation of the terrestrial vertebrate fauna. Factors such as phylogeny, habitat, ambient temperature, food intake, latitude, altitude, climate, season, body size, shape, level of development, degree of activity, sex and age variably determine the metabolic rate. The morphological complexities, refinements and respiratory efficiencies of the gas exchangers correlate with the metabolic capacities of animal taxa (e.g. Hughes and Morgan 1973; Dubach 1981; Gehr et al. 1981; Maina 1989b, 1996; Maina et al. 1989a). Furthermore, the metabolic capacities adequately reflect the phylogenetic level of evolution attained by different animal taxa. Dedicated biological engineering of the gas exchangers is hence axiomatic. The evolution of energetic animals paralleled development of more efficient gas exchangers. For reasons ranging from availability of O_2 to cost of breathing, for example, conceivably, the metabolic rates of the endotherms could not have been satisfied except by air breathing and even then only by gas exchangers ventilated by suctional breathing. Aerobic metabolism offers an efficient mechanism of energy production that is requisite for building, servicing and maintaining the infrastructural integrity of body tissues with high cell turnover. Fermentative (anaerobic) metabolism is an inefficient process of energy production. Much of the energy is left secured in the molecular bonds of alcohols and organic acids, products that must be eliminated before they accumulate to toxic levels. A molecule of glucose, for example, yields only 2 molecules of ATP that contain about 15 kcal of energy compared with 36 ATP molecules (equivalent to 263 kcal of energy) that are produced in aerobic metabolism. Notwithstanding the cost of uptake and delivery of molecular O_2 to the tissues, the benefits that arose from utilizing it for energy production far outweighed the dangers that it posed to life.

3 Themes and Principles in the Design of Gas Exchangers

When something is designed, decisions are necessarily made. Deliberate progress cannot proceed without choices – as to whether a part goes to the right or left of another part, whether a component is larger or smaller. In some cases a clear historical record documents for us the evolutionary design process that brought us the thing that we contemplate. Looking into that record in detail can help us understand why certain things are the way they are and, perhaps more importantly, help us understand how things in general come to take the forms that they do.

Petroski (2000)

Biology is now sufficiently mathematically emancipated from a purely observational, descriptive and classificatory scientific discipline to a quantitative (exact) discipline involving complex theories and models. Morphometric analysis and comparison of biological systems that share common designs offers a powerful approach in understanding the basis of evolutionary and adaptational advances in biology. Because natural selection exacts survival-specific (i.e. adaptive) features, biologists strive to explain the occurrence of analogous designs by arguing their consequential functional utility. Designs that are convergently acquired, i.e. those structures that have been 'preferred and mass produced', should bespeak optimal constructions.

Depending on the respiratory medium utilized, gas exchangers have basically evolved as evaginations (out-foldings) or invaginations (in-tuckings = cavitations = sacculations) from the surface of the body (Fig. 1). The gills, the primeval respiratory organs, fall into the first category, whereas the lungs (including the tracheal system of insects and the book lungs of the arthropods) fit into the second category. For successful transition from aquatic to terrestrial life, invagination of the gas exchangers was imperative for efficient water conservation on the desiccating terra firma. Moreover, internalization of the gas exchangers provided better protection from trauma and granted appropriate adjustments of inspired air, especially regarding moisture content and removal of pathogenic microorganisms and harmful particulate matter. An extensive respiratory surface area could be 'built' in a limited space through internal subdivision of the gas exchanger. In the adult human lung, for example, there are about 300 million alveoli of an average diameter of 250 μm (e.g. Weibel 1963, 1973) and a respiratory surface area of about 150 m^2 that is packed in only about 4 l of lung volume (e.g. Gehr et al. 1978). Theoretically, if the human lung was evaginated, i.e. if it hung out like the gills of fish and was thus freely exposed to the atmosphere, even in a moderately desiccating environment, the water loss would be about 500 l/day (McCutcheon 1964): That would be about 1,000 times more than the normal loss. Depending on lifestyle and habitat occupied, in insects, a balance between O_2 demand, water loss and spiracular opening frequency is imperative.

The internalization of the gas exchangers was accompanied by certain functional limitations. Although the evaginated gas exchangers (Fig. 1B–D) can be ventilated unidirectionally and continuously and even passively by forward movement, ram ventilation, the 'dead-ended' invaginated gas exchangers (Fig. 1D–F) can only be ventilated periodically (bidirectionally = in and out = tidally), a mechanism inferior to that compatible with the externalized gills. Moreover, the construction of the gills can support the highly efficacious counter-current gas exchange system in which the gills are perfused with blood and ventilated with water in opposite directions (Fig. 3C). In lungs, only the inefficient uniform pool arrangement is feasible (Fig. 3D). The internalized gas exchangers cannot fully exploit the high ambient Po_2 because the inspired air is 'diluted' by the stale air that is permanently resident in the upper conducting passages. In a resting human being, for example, where the dead-space is about 140 cm^3, about 28% of the 500 cm^3 of the tidal volume does not reach the alveolar level (e.g. West 1974). Furthermore, the partial pressure of O_2 is reduced from 21 kPa in the atmosphere to about 13 kPa at the alveolar level. On the plus side, invaginated gas exchangers can allow creation of important respiratory micromilieus. In the vertebrate lung, for example, the alveolar Po_2 is lower and the Pco_2 higher than in the atmospheric air. The high alveolar Pco_2 is used in the bicarbonate (HCO_3-) ion-mediated buffer system for pH regulation. Such microenvironments cannot form in the evaginated gas exchangers, where other ways and means of pH adjustment had to develop.

As Stebbins (1984) observed 'the only law that holds without exception in biology is that exceptions exist for every law', in the avian lung, an invaginated gas exchanger, by isolating the lung from the air sacs (Fig. 4) and shunting the inspired air to the caudal air sacs, the periparabronchial gas exchange tissue (Figs. 5, 6, 17, 114) is ventilated continuously and unidirectionally (e.g. Scheid 1979; Maina 1996), just as occurs in the evaginated gas exchangers! This provides the highly efficacious cross-current gas exchange system (Fig. 3E). The bimodal breathers, an important taxon that thrives at the air-water interface, extract O_2 from water using gills and air through accessory respiratory organs. The African catfish, *Clarias mossambicus*, for example, has well-developed gills that it uses in water and suprabranchial chamber membranes (Figs. 19, 116) and labyrinthine organs (Figs. 48, 124) that are very well vascularized as accessory respiratory (air breathing) organs (e.g. Maina and Maloiy 1986) when the fish is exposed to hypoxia. The labyrinthine organs arise from the gill arches and protrude into the suprabranchial chambers, concavities that extend above the branchial arches. The respiratory scheme that has evolved in animals such as *Clarias mossambicus* is shown in Fig. 1D. The lungfishes, *Protopterus aethiopicus* (Maina and Maloiy 1985; Maina 1987a) and *Lepidosiren paradoxa* (e.g. Hughes and Weibel 1976) have vestigeal gills and elaborate lungs (Figs. 21, 22, 27, 29): their respiratory stratagem falls in category 1 C.

The respiratory surface area of a human lung is about 100 times more than that of the body. Extensive surface area and thin tissue barrier are necessary parameters for efficient gas exchange. They are reached at a real cost and risk to the very survival of an organism. Development of various lines of defence was necessary to ensure the integrity of a vast surface across which the body is exposed to the external environment. Mucus and ciliated cells entrap particulate matter in the upper respiratory passages (Figs. 9, 10), and macrophages (Figs. 11, 16) phagocytose pathogenic microorganisms. In mammals,for example, the defense mechanism is so efficient that in the absence of a pulmonary disease, the respiratory system is virtually sterile below the larynx. Pulmonary macrophages are relatively few in the amphibian lungs and probably in all those of the ectotherms (e.g. Maina, 1989c). The numerical density of the pulmonary macrophages correlates with the level of environment pollution. Compared with mammals, birds have relatively fewer surface (free) macrophages (e.g. Maina and Cowley 1998). Secondary defence lines such as phagocytic atrial and infundibular epithelium (e.g. Scheuermann et al. 1997) and intravascular macrophages (Maina and Cowley 1998) have developed. In the honeybee, *Apis mellifera*, invasion of the tracheal system by parasitic mites (*Acarapis woodi*) reduces metabolic rate in hypoxic air: The diffusion of O_2 to the tissue is impaired (Harrison et al. 2001).

4 Functional Requirements and Design of Gas Exchangers

I have become convinced not only that structure determines function, but that functional demand also determines structural design, be it through evolution or by modulation of design features.

Weibel (1984)

In nature, the aphorism that 'necessity is the mother of invention' has been pertinent to the elaboration of novel biological designs, as it has been to the advancement of human technology. As in architectural and human engineering designs, fabrication of biological structures entails exploitation of the physical properties (e.g. strength and reliability) of the constitutive components. Given specific quantities and kinds of structural materials, infinitely many types of constructions are theoretically possible. However, notwithstanding this possibility, in biology, only a calculable number of design forms eventuate. Constructional, phylogenetic, developmental, functional and ecological constraints prescribe particular phenotypes. These outcomes must be the most optimal ones under the particular circumstances in which they form and given the needs that they are programmed to fulfill. Considering the remarkable morphological diversity of extant animals, the manifested structural (Fig. 1) and functional (Fig. 3) designs of the gas exchangers represent a very narrow range of the forms that may have evolved and far fewer than those that are theoretically possible. Through remarkable convergence, only a few different designs have developed. This conceivably indicates the importance of respiration as a physiological process for the sustenance of life and hence the superlative endeavours to optimize it. About 99.99% of all the animal species that have ever evolved on Earth are now extinct (e.g. Pough et al. 1989). Notwithstanding the particular cause of their demise, the designs of such animals must be considered to have been failed experiments.

In its broadest context, respiration is a process in which molecular O_2 is captured from an external fluid medium and made available to the tissue cells of an organism for energy production. It involves a complex arsenal of structural compartments and well-coordinated physiological, biochemical and behavioural processes. Through appropriate sizing, geometric patterning and spatial arrangement of the constitutive structural components, over the evolutionary continuum, the designs of the gas exchangers have been carefully crafted by natural selection for optimal procurement and delivery of O_2 to the tissue cells. This has entailed progressive refinements and transformations of the default gas exchangers. Modelling biological structures by changing proportions, dimensions, positions, shapes, sizes and arrangement of the integral components creates different or additional polarities and capacities that grant new functional capacities and hence ecological opportunities.

In the gas exchangers, adaptive changes have occurred either by attrition (deconstruction) and replacement of the default organ(s) with totally new one(s) or through production of auxiliary respiratory organ(s) at different site(s) (Maina 1998a). During the transition from water to land and conversion from breathing water to air, respiratory media of remarkably different physical attributes, deconstruction of the gills and construction of the lungs occurred. The gills gradually regressed as the lungs developed, largely as diverticulae of the foregut. Transitional (bimodal) breathers have gills and accessory respiratory organs that procure O_2 from water and air, respectively (e.g. Maina and Maloiy 1986) (Fig. 19). As has often occurred in the gas exchangers and elsewhere in biology, rather than entirely relinquishing an evolved structure, a totally different role has been assigned. The surfactant lining, for example, is thought to have evolved in the ancestral piscine lung for physical and chemical protection of the epithelial surface (e.g. Liem 1987). With the development of complex intensely subdivided lungs, the surfactant adopted the very important role of surface tension alleviation. Compared with other organs, the structural constitution of the gas exchangers is exceptional. There are no tissues or cells that are strictly specific or universal to them, for example, like hepatocytes are to the liver, osteocytes to bone, podocytes to the kidney and neurones to nervous tissue. The diffuse group of cells usually called pneumocytes has very few morphological features that are absolute to the respiratory role. The cell membrane of the unicellular organisms is for all intents and purposes a functional gas exchanger (Fig. 1A). In the lower vertebrates, for example, the lungfish (Dipnoi) (e.g. Maina 1987a), amphibians (e.g. Maina and Maloiy 1988; Goniakowski-Witalinska 1995) and certain reptiles (e.g. Pastor 1995), the pneumocytes have not fully differentiated into type I and II cells. This has only fully occurred in the lungs of birds and mammals (e.g. Corral 1995) (Figs. 30, 33, 36, 64).

When need and circumstances have enforced it, in rare cases of reverse evolution, gas exchangers have been remodelled through deconstruction and past simpler designs have been adopted. Some lineages of previously air-breathing slugs have, for example, reverted to water breathing using gill-like organs (ctenidia) developing in the antecedent lung (e.g. Cheatum 1934). Nature is particularly conservative. Attrition of the gas exchangers must have occurred only as extreme measures, i.e. when morphological transformation was not possible or was inadequate to accommodate external selective pressure. Considering the time taken and the cost incurred in founding organs and organ systems, this should make good practical sense. Insects that reverted to aquatic life, for example, retained their tracheal respiratory system, which served them very well on land: They have adopted plastron respiration (e.g. Hinton 1966) by which they carry air in the form of bubbles attached to their bodies for use during their underwater sojourn. Such insects are strictly air breathers while securing certain benefits of subsisting under water.

Life's tenacity for survival is remarkable. Anoxia (lack of O_2) or even lack of sunlight, the primary energy source in the biosphere, does not foreordain absence of life. In symbiotic associations between invertebrates and chemoautotrophic endosymbiotic bacteria, sulphur compounds have been used as sources of energy. Energy in the form of ATP is used in carbon fixation through the Calvin–Benson cycle, providing nutriments for the endosymbionts (e.g. Bernhard et al. 2000). Whether between animals, between animals and plants or between plants, symbiotic coexistence is enforced by extreme circumstances. Animals with intracellular sulphur-metabolizing endosymbionts are found in a wide range of anoxic environments ranging from deep-sea hydrothermal vents to sewage outflows, mangrove swamps and stagnant anoxic ocean basins (e.g. Anderson et al. 1990; Maina and Maloiy 1998; Maina et al. 1998; Bernhard et al. 2000). Hydrogen sulphide (H_2S) occurs in high concentration in such habitats. It is extremely toxic even at very low molar concentrations. A level greater than 1 ppm is lethal to most organisms. Binding to the heme of mitochondrial cytochrome *c* oxidase, much as cyanide does, the gas inhibits several heavy metal-containing enzymes, preventing O_2 transport by the haemoglobin (e.g. Somero et al. 1989). Certain animals have, however, developed the capacity to tolerate H_2S and even to utilizing it for metabolic processes: Water and various sulphates are the end products of metabolism. Amazingly, taxa such as the Phylum Gnathostomulida, the turbellarian families Solenofilomorphidae and Retronectidae have adapted so well to H_2S-rich biomes that they are virtually confined to them. Endosymbiotic bacteria occur, for example, in the gills of gutless cram, *Solemya reidi* (e.g. Anderson et al. 1990), the epithelial cells of the hindgut of *Urechis caupo* (e.g. Menon and Arp 1992) and the respiratory groove ('lung') of the swampworm, *Alma emini* (Figs. 37–39), a glossoscolecid worm that lives in the water-logged tropical swamps of East Africa (Maina and Maloiy 1998; Maina et al. 1998). In *Alma*, electron-dense intracytoplasmic granules (Figs. 40, 42, 43, 44) of the cells lining the respiratory groove were shown by X-ray microanalysis to be metabolites of sulphur compounds (Maina and Maloiy 1998). It was hypothesized by Mangum et al. (1975) that the 'lung' is used to carry air/water underground, serving as an O_2 reserve.

The remarkable physical and biochemical differences between water and air, for example, viscosity, thermal capacity and solubility and diffusivity of the respiratory gases (e.g. Randall et al. 1981; Dejours 1988) account for the fundamental morphological differences between the gills and the lungs. The distinctions are so consequential that those respiratory 'devices' that are efficient in water perform dismally in air and vice versa. In water, the gills are highly effective in O_2 acquisition, but in air, the closely packed leaf-like gill filaments and secondary lamellae (Figs. 46, 47, 49, 52, 54, 66, 111) soon dry up and become impermeable to gases. Moreover, after withdrawal of the physical support provided by water, the secondary lamellae cohere under surface tension and collapse under their own weight, much as the shore weeds flop at low tide. The respiratory surface area is greatly diminished, and large diffusional dead-air spaces are created between the secondary lamellae, reducing O_2 uptake. The animal succumbs to asphyxia although ironically exposed to a medium (air) that is richer in O_2.

Among the air breathers, except for the avian lung and the insectan tracheal system where the terminal gas exchange components, the air capillaries (Figs. 7, 8, 57–60, 62, 65, 113) and the trachea and tracheoles (Figs. 67–72, 74, 76, 77, 102, 103), respectively, are tubular rigid air conduits, in the internalized compliant lungs, for example, those of the amphibians, reptiles and mammals, foam-like terminal air cells surrounded by a moist surfactant-lined epithelium occur (Figs. 15, 21–24, 26, 27, 78–85, 104, 105, 108, 109). All conditions being the same, in practical terms, it requires more energy to inflate a lung with small terminal air spaces than one with larger air spaces. It therefore follows that although internal subdivision and compartmentalization produces an extensive respiratory surface, it obligates more energy to ventilate mechanically. Accordingly, to optimize the function of a gas exchanger, it is requisite that compromise between design, construction, operation cost and infrastructural outlay and maintenance be appropriately transacted. The low-metabolism ectotherms have evolved lungs that range from smooth-surfaced ones to those with large terminal air spaces (Figs. 15, 21–24, 26, 27, 80–83, 104–106, 108, 109). The mammalian lungs have developed a honeycombed parenchyma (Figs. 45, 78, 79, 84) with somewhat hexagonal to spherical alveoli (Figs. 78, 79, 84, 85, 110). The smallest alveolar diameter (~50 μm) in a mammalian lung was reported in an unnamed bat lung by Tenney and Remmers (1963). Birds have evolved a virtually rigid, noncompliant lung (Figs. 4, 5) with remarkably small terminal gas exchange components, the air capillaries (Figs. 7, 8, 57–60, 62, 65), the diameters of which range from 8 to 20 μm (e.g. Duncker 1971, 1974; Maina 1982; Maina and Nathaniel 2001). In snakes (Ophidia) and caecilians (Apoda = Gymnophiona), vertebrate taxa that have adopted a serpentine (cylindrical) body form that is thin and highly restrictive, there is a preponderance of single lungs (e.g. Maina and Maloiy 1988) (Fig. 25). In the African caecilian *Bourengerula taitanus* (Maina and Maloiy 1988), the tubular right lung is subdivided by septa that extend from two robust diametrically placed trabeculae that offer mechanical support and maintain patency to the lung (Fig. 26): Smooth muscle collagen and elastic tissue elements preponderate in the septa and trabeculae (Figs. 123, 128). The remarkable elasticity of the lung may explain how the South American aquatic caecilian *Typhlonectes natans* is able to empty its lungs in a single exhalation (Gardner et al. 2000). In the African lungfish *Protopterus aethiopicus*, the air duct is eccentrically located (Maina, 1987a; Maina and Maloiy 1985) (Fig. 21). Abundant smooth muscle and elastic tissue confer the tractability necessary for ventilation of elongate lungs (e.g. Stark-Vancs et al. 1984) (Figs. 89, 90). In *Pipa pipa* (Salentia: Pipidae), cartilaginous plates located in the first order of the septal walls provide internal support to the lung (Marcus 1937). In the caecilians *Chthonerpeton indistinctum* and *Ichthyophis paucisulcus*, tiny aggregates of cartilage cells occur in the proximal part of the lung (Welsch 1981). Recently, Maina and Africa (2000) described a highly vascularized epithelial swelling in the extrapulmonary primary bronchus of the lung of the domestic fowl, *Gallus gallus* var. *domesticus*, just ahead of the origin of the first medioventral secondary bronchus (Figs. 91, 92), a structure that corresponds to the segmentum accerelans that was presumed by Wang et al. (1992) through physiological experimentation. Together with other anatomical features including the angulation of the medioventral secondary bronchi (Figs. 5, 93, 94), the narrowing may be involved in inspiratory aerodynamic valving, namely, the rectification of the inspired air past the openings of the medioventral secondary bronchi (Wang et al. 1992). A segmentum accerelans is, however, lacking in the lung of the ostrich, *Struthio camelus* (Maina and Nathaniel 2001), suggesting that the underlying structures that determine the flow of the inspired air may differ among different groups of birds.

5 Respiratory Media Versus Design of Gas Exchangers

An important goal in biology is to uncover the fundamental design principles that provide the common underlying structure and function in all cells and microorganisms.

Joeng et al. (2000)

Of the three states of matter, namely, solids, liquids and gases, only fluids (liquids and gases) are atomically and molecularly suitably configured to contain molecular O_2 and to submit it to convective transport for delivery to the respiratory site. Water, a liquid over the biological range of temperature and pressure, and air, a gas under similar conditions, are the only two naturally occurring respirable fluids. The biophysical properties of water and air have in general greatly influenced the life patterns and the body forms of animals that subsist in the two media. Regarding gas exchangers, the differences have so fundamentally affected the respiratory processes that the basic mechanisms for obtaining O_2 and for eliminating CO_2 that are efficient in water often fail in air and vice versa. Compared with air, water is a more exacting respiratory medium. In saturated water, at 20°C, 1 ml of O_2 is contained in 200 g of water, whereas 1 ml of O_2 occurs in 5 ml (7 g) of air. The rate of diffusion of O_2 in water (3.3×10^{-5} cm^2 s^{-1}) is lower by a factor of 105 compared with that in air (1.98×10^{-1} cm^2 s^{-1}). Although the capacitance coefficient of O_2 (i.e. the increase of concentration per unit rise in the partial pressure) in water is only 1.82 nmol min^{-1} 0.133 kPa^{-1}, that in air is much higher (54.74 nmol min^{-1} 0.133 kPa^{-1}) (Dejours 1988). Furthermore, with respect to density and viscosity, water is about 800 times heavier and 50 times more viscous than air. Owing to these differences, to extract the same amount of O_2 from water and air, all other factors being equal, water breathers have to expend more energy than air breathers. An octopus, for example, ventilates 17 l of water for each millimole of O_2 consumed (e.g. Dejours et al. 1970). Water breathers have evolved within the constraints of an O_2-deficient and more viscous medium. The cross-current presentation (Sect. 9) and double capillary exposition (Sect. 12) of blood and water in the gills enhance O_2 uptake.

Except for the unicellular microorganisms and the small tracheates that obtain O_2 entirely by passive diffusion, in complex multicellular life, a respiratory medium in which O_2 is 'dissolved' is transported by bulk flow to the respiratory site, a process that entails energy expenditure. In fish, the cost of breathing water per unit O_2 uptake is estimated to range from 0.5% to 10% during routine ventilation and may rise to as high as 70% in exercise (e.g. Rantin et al. 1992). The vertebrate lung is estimated to utilize between 5% and 10% of the O_2 uptake (e.g. Slonim and Hamilton 1971). Because the mammalian lung is a passive organ that tractably reacts to the expansion of the rib cage and the diaphragm, much of the O_2 consumed by the respiratory organ must go to meet the metabolic needs of the structural components. Since they comprise living tissue, by default, gas exchangers must accede to a 'conflict of interest': they must consume O_2 before delivering the excess to the tissue cells. To minimize the amount of O_2 utilized in this way, minimal tissue must be committed to the construction of the gas exchangers without compromising their structural integrity. Cellular and supportive elements are qualitatively and quantitatively reduced and appropriately distributed to optimize respiratory and nonrespiratory functions. Such a compromise is necessary because, although gas exchange by passive diffusion is enhanced by minimal tissue infrastructural commitment, a critical amount is needed to support the metabolic (nonrespiratory) functions. By having the highly attenuated type I cells (Figs. 30, 86, 95, 96–99, 125, 126) lining most of the respiratory surface, the mammalian and avian lungs have developed an extremely thin blood-gas (tissue) barrier (e.g. Gehr et al. 1981; Maina 1989a, 2000c,d; Maina et al. 1989a). In the human lung, the type I cells constitute only 8.3% of the total number of pneumocytes but cover as much as 93% of the alveolar surface (Crapo et al. 1983).

In the fish gills, gas exchange occurs across the flap-like secondary lamellae (Figs. 46, 52, 54, 55, 107) while the nonrespiratory functions take place in the mucous cells and chloride cells (Figs. 12, 117, 119) that are found in the elaborate primary epithelium (Figs. 51, 52, 54) that is found over the gill filament and in the interlamellar spaces (e.g. Maina 1990a, 1991). In the avian lung, the type II cells and most of the perikarya of the type I pneumocytes are confined to the atrial and infundibula regions of the periparabronchial tissue (Scheuermann et al. 1997; Maina and Cowley 1998) (Figs. 6, 17, 18, 64, 114). Thin cytoplasmic extensions of the type I cells continue to the air capillary level (Figs. 97, 98).

6 Compromises and Trade-Offs in the Design of Gas Exchangers

The quality of a system depends on the quality of the components which form it, as well as the excellence of its organization. On the other hand, systems can improve the performance of components such as mechanisms and structures.

French (1988)

Unlike the human-made devices that are normally contrived to perform a specific task, biological structures are generally dynamic, composite, multifunctional entities. In their development, genetic programming engages various constraining factors including the available space in which they form. In mammals, for example, the maximal volume of the lung is determined by that of the thoracic cavity. In turn, the volume of the lung sets the dimensions of the components that constitute it. In some amphibian groups, the lungs play various roles that include hydrostatic adjustment, sexual display and sensory perception. The lung is an important source of pharmacologically active agents (e.g. Bakhle 1975), constitutes an important organ for defence and is involved in clearance of mucus and electrolyte transport. Regarding the endocrine and metabolic roles of the lung, Bakhle (1975) declared that 'instead of referring to pharmacokinetics as one of the functions of the lung, we ought to refer to gas exchange as one of the nonpharmacokinetic functions of the lung'. The exceptional diversity of cells in the lung (e.g. Weibel 1984) provides the necessary receptor sites for perception and execution of its various functions. During the development of the mammalian lung, as many as 20 different cell types are in place as early as the 14th week of gestation (e.g. Kauffman 1980). This affords protection as well as control of endogenous and exogenous biologically active chemical factors. In the dog and the rat, pulmonary capillary endothelial cells constitute as much as 51% of the total population of cells (Crapo et al. 1983). The endothelial cells of the pulmonary blood vessels have abundant micropinocytotic vesicles (Figs. 89, 90, 95, 97, 98, 134) that are involved in the degradation, transformation, interaction and biosynthesis of the macromolecules that the lung has affinity for as well as transendothelial transfer of some across the blood–gas (tissue) barrier. As in the lung, a multiplicity of functions occurs in the gills. These include osmoregulation, acid-base balance, modification of plasma hormones and elimination of products of nitrogen metabolism. To optimize these functions, refinements particularly regarding density and location of the structural components are evident in fish gills. Gas exchange occurs across the thin, flat cells that overlie the secondary lamellae (Fig. 55), whereas the metabolic functions occur in the more elaborate primary epithelium (Maina 1990a, 1991) (Figs. 51, 52, 54). Owing to their multifunctionality, the designs of gas exchangers must not be looked at only from the narrow perspective of their respiratory function but rather from a holistic perspective, i.e. from the viewpoint of their diverse roles.

Compromise design is intrinsic to the fabrication of the gas exchangers. The modern gas exchangers are refined products of a long trial-and-error process. Painstakingly, the working capacities of the respiratory organs were sized and regulated for optimal function to meet the demands placed on them by the lifestyle that animals pursued and the environment that they subsisted in. Inadequate designs were abandoned while the effective ones were adopted and conserved. The permutations involved in the development of the remarkably long trachea in birds presents an excellent example of the nature and extent of the compromises and trade-offs that occurred in the enterprise to achieve and preserve efficient gas exchange in the taxon. To attain flight, an energetically highly costly locomotory activity, but one that led to remarkable adaptive radiation and speciation in birds culminating in about 9,000 species, birds totally committed their forelimbs to that function. To compensate for the important roles that the forelimbs performed, birds developed a long, flexible neck (and with it a long trachea) for defence, procurement and manipulation of food, construction of nests and preening. In animals of similar body mass, a bird's trachea is 3 times longer than that of a mammal (e.g. Hinds and Calder 1971). To offset what should have occasioned high resistance to airflow, the trachea acquired a diameter that was 1.3 times greater than that of a mammal of comparable body mass. The consequence of this adjustment, namely, increased diameter of the trachea, resulted in an overall resistance equal to that of the much shorter trachea of a mammal of similar size. However, in acquiring a large tracheal volume and hence a large tracheal dead space that is 4.5 times greater than that of a mammal, this would have hampered respiratory function. This was craftily averted by development of a much lower respiratory frequency and high tidal volume (e.g. Bech and Johansen 1980): The respiratory rate of a bird is 0.32–0.42 times slower than that of a mammal of similar size.

7 Safety Factors in the Design of Gas Exchangers

Three guidelines of modern architecture are equally valid for biological forms: a building should be the realisation of clear, elegant structural ideas; it should be solid and economic; and it should be aesthetically harmonious with the environment.

Tributsch (1984)

The extent to which natural selection has produced concordance between the functional capacities of structural systems and their demands was termed 'symmorphosis' by Taylor and Weibel (1981). The concept has become a highly debatable and controversial subject in morphology, physiology and evolutionary biology (e.g. Weibel et al. 1991). As regards O_2 delivery systems (e.g. Weibel et al. 1991; Maina 1998b), it has been found that symmorphosis is tenable only as far as the system's components are quantitatively proportional to the maximal organismal functional requirements and loading. Since gas exchangers must meet the metabolic needs for O_2 by the body under shifting conditions, that is, rest to exercise, under thermal stress or under hypoxia, they must be structurally and functionally tractable. Highly trained human athletes and some elite animals such as the horse, the dog, the pronghorn antelope, *Antilocapra americana*, and the cheetah, *Acionyx jubatus*, can increase their O_2 consumption above rest by factors 20–30 times (e.g. Lindstedt et al. 1991).

The extent to which a biological system can contain excessive loading attests to its safety factor (= reserve capacity). Engineers define a 'safety factor' as the ratio of a component's capacity (or strength or performance) to the maximum expected load upon that component during operation'. Although in biology high safety factors should ensure survival, they inevitably obligate greater infrastructural outlay, that is, they call for greater commitment of resources for construction and cost of maintenance. Since the resources available to an animal are always finite, safety factors are instituted after thorough cost-benefit analysis. In biology, natural selection is the main factor that sets the magnitudes of safety factors. Different safety factors that allow various organs and organ systems to function under various excess capacities relative to normal functional needs have developed (e.g. Diamond 1998; Harrison et al. 2001). The fine line between safety factors and superfluosity, however, remains controversial (e.g. Weibel et al. 1998).

Diamond (1998) gave safety factors of 2 for the lung of a lazy cow and that of 1.25 for that of a fast, small dog. In the chicken egg, for example, the mean physiological diffusing capacities (DLo_2p) of the chorioallatois of 8.1×10^{-5} mlO_2 s^{-1} $mbar^{-1}$ (e.g. Piiper et al. 1980) compare with the morphometric diffusing capacity (DLo_2m) of O_2 of 8.5×10^{-5} mlO_2 s^{-1} $mbar^{-1}$ (Wangensteen and Weibel 1982). This indicates that gas transfer capacity across the eggshell is somewhat optimized. Wangensteen and

Weibel (1982) and Weibel (1984) attributed this to the ephemeral nature of the eggshell: it may not be justifiable to build functional reserves (at a cost) in a transient structure. In the human placenta, the DLo_2m ranges from 0.05 to 0.08 mlO_2 s^{-1} $mbar^{-1}$ (Mayhew et al. 1984). Estimated from the O_2 consumption of the uterus of a pregnant woman and the Po_2 in the maternal and fetal bloodstreams, Metcalfe et al. (1967) found the DLo_2p of the placenta to range from 0.014 to 0.018 mlO_2 s^{-1} $mbar^{-1}$. Calculations based on the diffusing capacities of CO_2 in pregnant women (Forster 1973), however, gave a higher value of 0.025 mlO_2 s^{-1} $mbar^{-1}$. On the basis of a mathematical model that assessed the effects of uterine contractions on placental O_2 exchange, Longo et al. (1969) arrived at a DLo_2p of 0.038 mlO_2 s^{-1} $mbar^{-1}$. The currently available data suggest that the DLo_2p of the human placenta ranges between 0.025 and 0.038 mlO_2 s^{-1} $mbar^{-1}$, a value that is lower than that of the DLo_2m by a factor of about 2. This difference can be deemed to be a safety factor. It is pertinent to point out that unlike the eggshell, the placenta, an equally transitory organ, preserves a sizable functional reserve during much of its life (e.g., Karsdorp et al. 1996). The maximal DLo_2m of the human lung of 2.38 mlO_2 s^{-1} $mbar^{-1}$ (Gehr et al. 1978) is about 30-fold the DLo_2m of the placenta. This offers a colossal functional reserve (safety factor) of at least about 63 times between the two gas exchangers. This should ensure O_2 transfer to the fetus under various limiting conditions and circumstances such as hypoxia and maternal pulmonary and placental pathology. In the mammalian lung, generally, the DLo_2m and the DLo_2p differ by a factor of 2 (e.g., Gehr et al. 1978; Weibel et al. 1983). The difference indicates a functional reserve that should be utilized during extreme circumstances. According to Hill et al. (1973), in the human lung, the alveolar-capillary barrier that in normal conditions is 0.62 µm thick (e.g., Gehr et al. 1978) would have to increase $4^{-1}0$ times before it becomes a limiting factor for end-capillary Po_2 equilibration. In the human lung, failure of the alveolar blood-capillary barrier has been associated with excessively high pulmonary capillary blood pressure (e.g. Severinghaus 1971). The pulmonary capillary blood pressure in the rabbit (Maarek and Grimbert 1994), the dog (Okada et al. 1992), and the human (Hellems et al. 1949), animals of remarkably different body sizes, is about 1.1 kPa. The alveolar wall in the lung of the rabbit *Oryctolagus cuniculus,* which is only 0.50 µm thick (e.g. Weibel 1973), fails above a transmural blood pressure of 3.3 kPa (e.g. Costello et al. 1992): This offers a safety factor of 3. In the dog, the resistance in a segment of an alveolar blood capillary remains stable even after large changes in the transmural pressure (e.g. Okada et al. 1992). In horses, where the pulmonary capillary blood pressure increases from 2.4 kPa at rest to about 6 kPa during exercise (Sinha et al. 1996), the capillary transmural blood pressure that causes stress failure of the blood capillaries ranges between 10 and 13 kPa (Birks et al. 1997), giving a safety factor of about 2.

Although the safety margin of operation of O_2 delivery by the tracheal system in insects is very high (e.g. Joss et al. 1997), it can be completely eroded during flight (e.g. Harrison and Lighton 1998). In the insectan tracheal system, gas exchange and hence flight capacity should ultimately be limited by the diameter of the air conduits, the trachea and tracheoles (Figs. 14, 67–72, 74, 76, 77, 102, 103). Large trachea occupy extra space and may physically interfere with muscle contraction. Furthermore, large trachea offer a greater diffusional distance of O_2 to the tissue cells. Thus optimization of the tracheal system for gas transfer should entail reduction in tracheal diameter. The mean free path of O_2 molecules ($MFPO_2$) in air, namely, the average distance an oxygen molecule travels before colliding with another, is about 0.008 μm (e.g. Weis-Fogh 1964b; Pickard 1974). Theoretically, in tracheoles narrower than the $MFPO_2$, the effective diffusion would be reduced. The $MFPO_2$ must therefore set the limit for the smallest functionally competent tracheoles. On average, the minimum tracheal diameter is about 6–10 times the $MFPO_2$ (e.g. Wigglesworth 1974). The ratio of minimum tracheal diameter to the $MFPO_2$ should indicate a safety margin of operation in a tracheal gas exchanger. In highly metabolically active tissues such as the flight muscle, however, the minimum tracheal diameter approximates to the $MFPO_2$. Infestation of the honeybee, *Apis mellifera*, by a tracheal mite, *Acarapis woodi*, a condition that reduces the diffusion and hence the delivery of O_2 to the tissues, lowers the safety margin of operation (Harrison et al. 2001): The mortality of bees in a hive (estimated as the number of bees that fail to return from feeding) correlates with the level of infestation in a colony: In 10% O_2 atmospheres, flight metabolic rate is reduced by 20% by moderate mite infection and by 40% by severe mite infection. All factors being the same, with the diameters of the air capillaries of the avian lung [where air flow occurs by diffusion (e.g. Scheid 1978)] ranging from 8 to 10 μm (Duncker 1974; Maina 1982; Maina and Nathaniel 2001), as regards the influence of the $MFPO_2$ on lung function, the safety is greater by a factor ranging from 1,000 to 2,500 times.

8 Malleability, Tractability and Performance of Gas Exchangers

Are organisms like liquid droplets, infinitely malleable by the changing forces of evolution, or do they contain a "frozen core" – the Bauplan, or body design, which remains little changed under the varying adaptive pressures a lineage encounters during its history?

Wagner (1998)

Genetic breeding and domestication of animals, human manipulations, practices and interventions that were termed 'directed evolution' by Joyce (1992), have induced incontrovertible unbalanced designs and performances of various organs and organ systems particularly upon the gas exchangers. The horse, dog and chicken are the foremost animals that have been domesticated and severely genetically exploited for various properties and features such as speed, temperament, weight gain and egg production. The domestic fowl, *Gallus gallus* var. *domesticus*, was domesticated from the wild jungle fowl, *Gallus gallus*, of South East Asia some 8,000 years ago. During the period that has elapsed, at least some 40 different breeds of commercial importance have been produced. Although better husbandry and feeding protocol may contribute to this, in the late 1940s, broilers took about 90 days to reach a slaughter body mass of 1.8 kg, in 1960 it took 70 days for a table bird to reach a similar live weight, whereas in the 1980s it took only 40 days (e.g. Gyles 1989). Presently, broilers reach a body mass of 2.5 kg in less than 40 days (e.g. Ross Breeders 1999). Most of this progress has arisen from a marked growth rate during the first 2 weeks after hatching (Ricklefs 1985).

In the process of directed (enforced) productivity/performance through intense genetic selection in total disregard of the importance of the necessary adjustments of the supporting structures, the functional integrity of certain organs and organ systems seems to have been severely compromised. The growth and productivity rates of chickens in particular appear to be approaching a zenith (e.g. Konarzewski et al. 2000). Deaths from aortic rupture (e.g. Carlson 1960) and vascular pathology (e.g. Silversides et al. 1997) now constitute as much as 10% of the total losses in the turkey industry. A worldwide increase in ascites in young broilers, a syndrome associated with right ventricular hypertrophy was reported by, among others, Julian and Wilson (1986). Huchzermeyer et al. (1988), for example, suggested that hypoxia may cause pulmonary vasoconstriction with consequent pulmonary hypertension. Interestingly, the village free-ranging chickens are not susceptible to ascites (Pizarro et al. 1970).

Respiratory inadequacies arising from the poor morphometric features that typify the domesticated flightless birds (e.g. Duncker 1971, 1974; Abdalla et al. 1982; Maina and King 1982; Vidyadaran et al. 1990) may exacerbate their poor functional capacity and survival. The development of the lung of a turkey does not match the growth and increase in body mass (e.g. Timwood et al. 1987). For similar reasons, namely, intense selection for productivity in total disregard of the consequences for the affected organs and organ systems, the domestic fowl (particularly the males) are totally incapable of achieving maximum O_2 consumption on treadmill exercise (Brackenbury 1984).

For the domesticated pig, *Sus scrofa,* in the year 1800 it took 3 years for the animal to reach a body mass of 40 kg, in 1850 pigs reached 70 kg in the same duration, whereas today they attain an average of 100 kg in 5–6 months. The growth rate has caused disproportionality between heart and body masses. In the pig, the heart constitutes only 0.3% of the body mass compared with higher values of about 1% in the horse and the dog. Sudden death syndrome, a condition that may be caused by excessive strain on the heart, is a common cause of death in pigs. Maina and Gils (2001) attributed the morphometric disparities between pulmonary structural parameters of the bronchial, venous and arterial systems of the lung of the pig to genetic manipulation and the indolent lifestyle that the animal now leads in captivity. Regarding the respiratory performance of the horse, a highly genetically bred animal, Jones (1998) observed that the animal appears to have reached the peak of its exercise performance. Jones (1998) pointed out that for the last 70–150 years, the winning times of the British and American thoroughbred races have 'reached asymptotic nadirs'. Pulmonary haemorrhage induced by exercise affects more than 40%–80% of horses (e.g. West et al. 1993).

9 Presentations of Respiratory Media in Gas Exchangers

One approach to uncovering biological design principles is to ask what constraints they must obey. Apart from the laws of physics and chemistry, most constraints arise from evolution, which has selected particular solutions from a vast range of possible ones.

Hartwell et al. (1999)

From a morphological perspective, the efficiency of a gas exchanger is, on the whole, set both by the geometric arrangement and the intrinsic qualitative and quantitative refinements of the constitutive elements. The arrangement of the structural components determines the transport modalities and the presentations of the respiratory media, whereas enhancement of the structural components sets the level of disposition (i.e. the exposure) of the gas exchange media to each other. When the respiratory media flow in the same direction, the presentation is termed 'con-current' (Fig. 3A); when the media run in opposite directions, it is termed 'counter-current' (Fig. 3C); when they run in perpendicular directions it is designated 'cross-current' (Fig. 3E) and when the external medium is held steady over or against a gas exchanger (Fig. 3B) or the gas exchanger is ventilated with a medium of which the Po_2 is fairly uniform (Fig. 3D), the arrangements are termed 'infinite pool' and 'uniform pool', respectively (e.g. Piiper and Scheid 1992).

By appropriately fitting or approximating the design of a gas exchanger to one of the functional models, the deficiencies and the limitations inherent in a particular design can be assessed. In a counter-current system, within the gas exchanger, deoxygenated blood is continually exposed to highly oxygenated incoming ('inhaled') water: A partial pressure gradient and hence a flux of O_2 into the blood is sustained during the exposure of the respiratory media. The exceptional efficiency of the counter-current system of fish gills is necessary for survival in a medium deficient of O_2. The O_2 extraction ratios in the fish gills and those of the cephalopods range from 50% to 80% (e.g. Hanson and Johansen 1970). In the triggerfish, *Balistes capriscus*, O_2 extraction is as high as 92% (Hughes 1967). When the direction of water flow in the gills is reversed, that is, establishing a con-current system, the O_2 extraction drops to well below 10%. In the cross-current system, however, a similar operation has little effect on the overall O_2 extraction (e.g. Scheid and Piiper 1972): Only the sequence of the blood capillary arterialization is altered (Figs. 3E, 100, 101). The cross-current exposure of the pulmonary capillary blood to the parabronchial air in the avian lung offers a highly efficient gas exchange system (Fig. 3E). Through an additive multi-capillary serial arterialization scheme, under hypoxic conditions and during exercise, the Po_2 in the arterial blood may surpass that in the end-expired air (e.g. Scheid and Piiper 1972).

In human-made devices, the efficiency of counter-current exchangers, be they of heat or gases, is determined by features such as surface area of exposure, thickness of the separation between the media, prevailing concentration or partial pressure gradient and optimum flow rates of the media. Chemical and mechanical engineers employ counter-current systems to enhance concentration gradients so as to promote mass transfer across systems. Various counter-current systems have evolved in biology. These include heat exchangers, for example, the carotid and ophthalmic rete; salt-concentrating systems, for example, kidneys and salt glands of birds; and gas-secreting organs, for example, the choroid rete and the rete mirabile of the swim (air) bladder of fish (e.g. Pelster and Scheid 1992; Maina et al. 1996a). In the epitheliochorial placentae of sheep and goat and the haemochorial placentae of the rabbit and the guinea-pig, a counter-current gas exchange system occurs between the fetal and maternal blood streams (e.g. Metcalfe et al. 1967).

The cell membrane of a simple microorganism and the skin of an amphibious animal offer an example of an infinite pool gas exchanger. The skin is, however, not a passive gas exchanger. Gas transfer can be adjusted through greater perfusion and 'ventilation' by physical movement in the surrounding medium. Blood flow to the mid-dorsal skin in amphibians is 1.8 times the blood flow to the ventral thoracic skin (Moalli et al. 1980). During transition from water to land and replacement of gills with lungs, the skin served as an important 'bridging organ' (e.g. Randall et al. 1981). In contemporary bimodal air-breathers, the gill-skin system removes about 76% of CO_2 while the lung eliminates the rest. In amphibious animals, O_2 uptake varies with the species, the kind of gas exchanger used and the habitat occupied. The skin of the tree frog, *Chiromantis petersi*, an extremely xerophilic amphibian, is highly vascularized (Maina 1989c) (Fig. 115)

10 Compartmentalization and Stratification of Gas Exchangers

To appreciate that nature does not necessarily have all the best ideas, we need only point to the wheel. Nevertheless, most of the problems of controlled motion and maneuverability have been explored and elegantly resolved in the living world.

Ball (2001)

Space is at a premium in the animal body. Internalized organs, especially those involved in exchange and transfer of materials and substances between the external environment and the body (e.g. the gas exchangers and the gastrointestinal system), must particularly be appropriately engineered. Various tree-like (= bifurcating = branching) structures, such as ducts of composite glands, blood and lymph vessels, neurones and bronchial and vascular trees of the respiratory organs (Figs. 118, 138–140), are manifest in animal morphology. Particularly in the gas exchangers, the respiratory surface area is an important factor in determining the conductance (flux) of the respiratory gases across the tissue barrier (e.g. Roughton and Forster 1957). At certain points in the their evolution, when metabolic demands dictated and environmental circumstances enforced it, augmentation of the respiratory surface area was requisite to promote respiratory efficiency. The respiratory surface area of a gas exchanger can be increased by overall enlargement of a gas exchanger and/or by increased internal partitioning (in case of invaginated gas exchangers-lungs) and greater and more intense structural stratification (in case of the evaginated gas exchangers-gills). Whereas a sphere of a volume of 1 cm^3 has a surface area of 4.8 cm^2, 1 cm^3 of the parenchyma tissue of the lung of the shrew, *Sorex minutus*, for example, has a profound surface area of 2,100 cm^2 (e.g. Gehr et al. 1980). In the bat lung, an extensive respiratory surface area has been produced mainly from increased lung volume (Maina et al. 1982; Maina and King, 1984; Maina et al. 1991), whereas in birds, it has occurred by more intense subdivision of the periparabronchial gas exchange tissue (Maina 1989b; Maina et al. 1989a). The mass-specific respiratory surface area of 138 cm^2 g^{-1} reported by Maina et al. (1982) in the lung of the epauletted fruit mbat, *Epomophorus wahlbergi*, is the highest value that has so far been found in a vertebrate lung. The extremely high respiratory surface area of 800 c^2 g^{-1} reported in a lung of an unnamed hummingbird by Stanislaus (1937) should be treated with caution. The value far exceeds that of 87 cm^2 g^{-1}, which was more reliably determined by Dubach (1981) in the lung of the violet-eared hummingbird, *Colibri coruscans*.

Increased internal subdivision of a gas exchanger produces small terminal gas exchange components. Except for the avian lung-air sac system and the insectan tracheal system, where the air capillaries (Fig. 7, 8, 57–62, 65) and the trachea and tracheoles (Figs. 67–72, 74, 76, 77, 102, 103), respectively, are tubular and rigid, in the expansible lungfish, amphibian, reptilian and mammalian lungs, the terminal gas exchange units fairly resemble foam air bubbles (Figs. 15, 21–24, 26, 27, 29, 45, 78–85, 104–106, 108, 109). In the lungs of the snakes *Dendroaspis polylepis* (Maina 1989d) and *Eryx colubrinus* (Maina et al. 1999), the faveoli present a beaded arrangement of discrete air cells that bifurcate peripherally into two or three terminal gas exchange units (Figs. 105, 106, 108, 109). In intensely compartmentalized compliant lungs (e.g. the mammalian lung), the small sizes of the terminal gas exchange components, the alveoli, generally make the gas exchanger susceptible to collapse owing to high surface tensional forces at the air-tissue interface. Moreover, high surface tension affects the fluid balance across the blood-gas (tissue) barrier. During the evolutionary continuity, trade-offs and compromises between the founding of efficient pulmonary design and optimal cost of operation of the gas exchangers were necessary.

The urodele salamanders (Plethodontidae) acquire all their O_2 needs across the skin from the cold, well-oxygenated water that they inhabit; they lack a functional lung (e.g. Piiper et al. 1976). The simplest vertebrate lungs are the smooth-walled, poorly vascularized saccular types that occur, for example, in *Proteus* (e.g. Hughes 1970). The moderately subdivided dipnoan lungs (Figs. 21, 22, 27, 29) are thought to be possible precursors of more complex lungs of the tetrapods (e.g. Gardner 1980) (Figs. 15, 17, 18, 26, 45, 57, 58, 78–85, 104–106, 108, 109). Among the air-breathing vertebrates, animals with low metabolism (e.g. ectotherms) have evolved lungs with large terminal air spaces (Figs. 15, 21–23, 26, 27, 29, 31, 80–83, 104–106, 108, 109), whereas the highly metabolically active endotherms, mammals (Figs. 45, 78, 79, 84, 85, 110) and birds (Figs. 7, 8, 57–62, 65, 113), have lungs with relatively small air spaces. Furthermore, the intricate gas exchangers have well-structured conducting systems (Figs. 5, 78, 118, 138–142). The lungs of archaic fish like the bichirs, for example, *Polypterus bichir*, are poorly internally subdivided (e.g. Budgett 1900). In caudates such as *Salamandra*, *Amphiuma*, *Megalobatrachus* and *Siren*, the lung is internally well subdivided (e.g. Stark-Vancs et al. 1984). The amphibian lungs are most elaborate in the Salentia, where peripherally situated shallow air cells confer a low surface-to-volume ratio (e.g. Smith and Rapson 1977). The single-chambered lungs with an edicular parenchyma are thought to obligate low energy expenditure for convective ventilation (e.g. Perry 1989). These include the lungs of the tree frog *Chiromantis petersi* (Maina 1989c) (Figs. 23, 24, 31), the caecilian *Boulengerula taitanus* (Maina and Maloiy 1988) (Figs. 25, 26), and the lungfish *Protopterus aethiopicus* (Maina and Maloiy 1985; Maina 1987a) (Figs. 21, 22, 27, 29). In the simple lung of the slug *Trichotoxon copleyi* (Maina 1989e) (Figs. 28, 133), haemolymphatic vessels bulge out into the air space, leaving deep crevices between the vessels. The design of the reptilian lungs ranges in complexity from the multicameral highly partitioned lung of the turtles, crocodiles, snakes and monitor lizards (Figs. 15, 80–83, 104–106, 108, 109) through the less elaborate (paucicameral) lungs of the chameleons and the iguanids to the simple, saccular, smooth walled, transparent (unicameral) lungs of, for example, the teju lizard, *Tupinambis nigropunctatus* (Perry 1989). Extrapulmonary saccular extensions that resemble the air sacs of the avian lung occur in lungs of the chameleons (Maina 1998a). The land-based chelonians have paucicameral lungs, that is, the lungs have two or three peripheral compartments that open into a central air space but an intrapulmonary bronchus is lacking (e.g. Perry 1989). The monitor lizards (varanids) present some of the highest levels of pulmonary complexity in the suborder Sauria (Maina et al. 1989b): The lungs are multichambered and intrapulmonary bifurcated bronchi occur (Fig. 15). The elongated lungs of the caecilians (Figs. 25, 26), the snakes (Ophidia) (Figs. 106, 108) and the amphisbaenids are remarkably heterogenous; they are generally divided into an anterior gas exchange region and a posterior air holding region. To a certain extent, the design of the ophidian (snake) lung is somewhat analogous to the lung-air sac system of birds, in which the gas exchange region (lung) is separated from the air sacs (Fig. 4) that serve as mechanical ventilators and air storage devices.

The compartmentalization of the reptilian lungs occurs by inward filling, that is, through a centripetal (= centralizing = compacting) growth. From the peripheral wall, septae advance towards the central air space, the unfilled spaces forming the air conduits. The mammalian and avian lungs, however, fill by outward, that is, centrifugal (= radiative = diffusive) growth of the airways. Although evidence is still tenuous, the bronchoalveolar lung of mammals and the parabronchial lung of birds are said to have evolved from transformation of the multicameral lung of reptiles (e.g. Perry 1989). Unique to other air-breathing vertebrates, the avian lung (Figs. 4, 5) is compact and virtually nonexpansile during the respiratory cycles; the volume changes by a mere 1.4% (e.g. Jones et al. 1985). Physical compression of the avian lung does not cause significant collapse of the air capillaries (Macklem et al. 1979). By synchronized action, the air sacs continually and unidirectionally ventilate the exchange tissue (e.g. Scheid 1979); the air sacs play no direct role in gas exchange.

Although birds have smaller lungs per unit body mass than mammals by a factor of 27% (Maina et al. 1989a), owing to their compactness, intense subdivision of the exchange tissue generated a respiratory surface area per unit body mass that is 15% greater than that of a mammal of similar body mass. The intensity of internal subdivision (Figs. 6-8, 17, 18, 57, 58, 60, 62, 65) is reflected in the very high surface density of the blood–gas (tissue) barrier (i.e. the surface area per unit volume of the parenchyma) (Maina 1989b, 2000b,c; Maina et al. 1989a), values that range from 236 to 389 $mm^2\ mm^{-3}$. On average, the surface density of the blood–gas (tissue) barrier of the avian lung is greater than that of mammals by a factor of 170 and 305%. The highest surface density of the blood–gas (tissue) barrier in mammals is that of 207 $mm^2\ mm^{-3}$ reported in the lung of a 2.88-g shrew, *Sorex minutus,* by Gehr et al. (1980). The attachment of the avian lung to the ribs (Figs. 4, 5), its virtual rigidity and the relegation of compliance to the air sacs allowed development of extremely narrow terminal gas exchange components, the air capillaries (Figs. 7, 8, 57, 58, 60, 62, 65, 113), because surface tension was not a constraining factor in the degree and extent of the internal subdivision of the gas exchange tissue and hence the size of the terminal gas exchange components. From the hummingbird to the ostrich, *Struthio camelus,* the air capillaries range from 8 to 20 ìm in diameter (e.g. Duncker 1974; Maina 1982; Maina and Nathaniel 2001). On the geometry and profusity of the pulmonary blood capillaries on the alveolar septa of the lung of the minute shrew, *Microsorex hoyi,* which weighs as little as 2.3 g, Weibel (1984) remarked that 'the honeycomb arrangement of the alveoli compared with the closely interdigitated air- and blood capillaries of the bird lung'; he added that 'the shrew lung may probably have been driven to the very limit of a functionally operable mammalian lung'. The same may apply on the lung of the 2-g Cuban bee humming bird, *Calypte helenae,* and that of the 1.6-g Thai bumblebee bat, *Craeonycteris thonglongyai.* Unfortunately, data on lungs of these interesting species are not presently available.

In fish gills, hundreds of gill filaments give rise to thousands of bilaterally situated secondary lamellae (Figs. 46, 52, 54, 107, 111), thin flattened flaps that serve as the respiratory units. In crustacean gills (Figs. 47, 49, 66), gill filaments offer extensive respiratory sites, whereas the book lungs of the spider are made up of layers of lamellae that are separated by spikes (Fig. 50); the similarities between these designs in phylogenically different animals is remarkable. In tracheates, a taxon of which the insects are the most highly developed, fine air conduits pervade the bodies. The trachea and the tracheoles (Figs. 14, 67–72, 74, 76, 77, 102, 103) are simple rigid hollow tubes strengthened by endocuticular helical or circular chitinous thickenings, the taenidia. Every individual cell in the body is directly or indirectly supplied with air from the atmosphere. Functionally, the tracheoles are analogous to the vertebrate blood capillaries. In insects, the respiratory and circulatory systems have been totally disengaged, with the later generally being of no practical consequence in gas exchange. In some cases, the tracheoles indent cells in a manner of a finger poked into a balloon (e.g. Wigglesworth 1974) (Fig. 68). The indenting tracheoles, however, remain ensheathed in their own tracheoblast as well as in the membrane of the indented cell (Miller 1974) (Figs. 70, 72). Depending on factors such as the stage of development and exercise capacity, the tracheal system constitutes as much as 50% of the volume of the body of an insect (e.g. Krogh 1941; Schmitz and Perry 1999): in the silkworm, *Bombyx mori*, there are about 1.5×10^6 tracheoles that are 1.5 m long. The trachea in the cossid moth, *Cossus*, and the water beetle, *Dystacus*, constitute 1.4% and 6%–10% of the body volume, respectively (Krogh 1941). Longitudinal and transverse trachea (Fig. 74) connect the air sacs (Figs. 73, 75). The air sacs increase the tidal volume by as much as 70% of the total tracheal air capacity (e.g. Wigglesworth 1974). As is the case in vertebrates, where blood capillary profusity correlates with the metabolic activity in tissues, the tracheal density is determined by the Po_2 and the metabolic intensities in different parts of the body (e.g. Miller 1974). In larval mealworms, *Tenebrio molitor*, hypoxia influences tracheal growth: An ambient Po_2 of about 10 kPa leads to wider trachea (e.g. Loudon 1989). In the flight muscles of the dragonfly, the tracheal density is so high that each muscle fibre is virtually suspended in air (Weis-Fogh 1964b). In such cases, the maximum diffusion distance is 10 μm and the tracheoles are no more than 20 μm apart. In Diptera and Hymenoptera (e.g. Miller 1974), the tracheoles indenting the muscle fibres lie within 2–3 μm of each other. In complete contrast to the bronchial system of the mammalian lung (Figs. 138, 141, 142), where between the principal and the terminal bronchi the velocity of air flow decreases by a factor of 700, the tracheal systems of certain insects, for example, *Rhodnius*, are the only known evolved air conduits in which the cumulative cross-sectional area remains more or less constant with length (e.g. Krogh 1941).

11 Separation of Respiratory Media in Gas Exchangers

The organism is a compromise. The result of natural selection is adequacy and not perfection.
Bennett (1987)

The thickness of the tissue partitioning between water/air and blood in part sets the conductance, that is, the diffusing capacity (Do_2) of a gas exchanger. Furthermore, Do_2 is determined by factors such as the respiratory surface area, the volume of the pulmonary capillary blood, the prevailing partial pressure gradient and the permeability of the tissue compartments that O_2 diffuses across (Fick's law). The Do_2 correlates directly with the surface area (S) and the permeation constants (K) and inversely with the thickness of the tissue barrier: $Do_2 = S \quad Kt^{-1}$. In the vertebrate air breathers, the thickness of the blood-egas (tissue) barrier generally decreases from amphibian, reptilian, mammalian to avian lungs. Owing to the thick blood-gas/water (tissue) barrier, which measures between 20 and 50 μm, cutaneous gas exchange is characterized by limitations of diffusion (e.g. Malvin 1988). Increased vascularity (Fig. 115) and perfusion may promote O_2 uptake, particularly during hypoxic episodes (e.g. Malvin and Hlastala 1986). In fish gills, a thick water-blood barrier (Hughes and Morgan 1973) (Figs. 54, 55) and the presence of an unstirred boundary water layer over the secondary lamellae may curtail O_2 transfer, producing significant diffusion limitations (e.g. Randall and Daxboeck 1984). Although generally thicker than the blood-gas (tissue) barrier of the lungs, the water-blood barrier in the gills of certain species of fish may be as thin as 0.2 μm in some parts (e.g. Hughes and Morgan 1973; Maina 1991). In mammals (e.g. Gehr et al. 1981) birds (e.g. Maina 1989b; Maina et al. 1989a) and even fish (e.g. Hughes and Morgan 1973), the thicknesses of the tissue barriers of the gas exchangers correlate with the metabolic needs as well as the environment in which they live. The tuna, *Katsuwonis pelamis*, one of the most energetic fish, has a water-eblood barrier that is only 0.6 μm thick (e.g. Hughes and Morgan 1973), and *Oreochromis alcalicus grahami*, a fish that lives in the hot, highly alkaline Lake Magadi of Kenya that is virtually hypoxic at night, has a barrier that is as thin as 0.2 μm (e.g. Maina et al. 1996b).

The pathway that O_2 traverses from the external environment to the tissue cells and ultimately into the mitochondria consists of a panoply of tissue components and spaces (Figs. 2, 54, 55, 97–99, 121) that differ in complexity and design. To promote the flux of the respiratory gases across a thin barrier, minimal tissue is located over the respiratory surface. In fish gills, for example, a thin secondary epithelium overlies the secondary lamellae (Figs. 54, 55), whereas the cellular elements such as the chloride (= mitochondria rich = ionocytes) cells and mucous cells are found in the composite primary epithelium that lines the gill filaments (Figs. 12, 51, 117, 119). The nonrespiratory functions of the gills such as osmoregulation and ammonia excretion occur in the primary epithelium, whereas gas exchange takes place strictly across the secondary epithelium; an epithelial cell and its basement membranes, an interstitial space and an endothelial cell with its basement membrane essentially constitute the water-blood barrier (Figs. 54, 55). By locating most of the cellular and connective tissue elements in the primary epithelium, the water-blood barrier becomes thinner. In fish (Figs. 52–56, 107, 111) and crustacean (Figs. 66, 121) gills, endothelial cells are arranged as struts that span the space between the parallel, diametrically located epithelial cell sheets. While slowing down the flow of blood and hence promoting gas exchange, the pillar cells render structural integrity to the secondary lamellae and gill filaments of the fish and crustacean gills, respectively. They prevent overdistension and mechanical damage of the vascular spaces under excessive intramural blood pressures. In the gill filaments of *Potamon niloticus* (Maina 1990b), prolongations of epithelial cells form sporadic zip-like pillar extensions (Fig. 121).

In the fish gills and the lungs of the lungfishes (Dipnoi), amphibians and reptiles (e.g. Hughes and Weibel, 1976; Maina 1987a; Maina and Maloiy 1985, 1988; Perry et al. 1989), an interstitial space occurs between the basal cell membranes of the epithelial and endothelial cells (Figs. 31, 55, 95, 128, 132, 134). The interstitial space may contain nerves (Fig. 134), lymphatic vessels and connective tissue elements such as collagen, elastic tissue and smooth muscle (Figs. 89, 90). In the blood-gas tissue barrier of the avian lung, an interstitial space is, however, lacking (Figs. 97, 98): The basal cell membranes of the epithelial and endothelial cells have fused into a common basement membrane. This has contributed to the remarkably thin blood-gas (tissue) barrier that characterizes the avian lung (e.g. Dubach 1981; Maina 1989b, 2000b, c; Maina et al. 1989a). In rare cases where air capillaries lie adjacent to each other (Figs. 58, 62, 126), a basement membrane is lacking. This indicates that, developmentally, the formation of the basement membrane is determined by the endothelial and not the epithelial cell. In the avian lung, the endothelium constitutes 67%, the basal lamina 21% and the epithelium 12% of the volume of the blood-gas (tissue) barrier (Maina and King 1982). The endothelial cells of the blood capillaries of the gas exchangers contain numerous micropinocytotic vesicles (Figs. 89, 90, 95, 97, 98, 134), and those of the avian lung are particularly highly sporadically attenuated (Figs. 97, 98, 126). The ratio between the arithmetic mean thicknesses (τt) and the harmonic mean thicknesses (τht) of the blood-gas (tissue) barrier indicates the degree and intensity of corrugation of the blood-gas (tissue) barrier (e.g. Maina and King 1982). In the avian lung, the mean of the minimum τht of the blood-gas (tissue) barrier is as small as 0.068 μm (Maina and King 1982) and the mean ratio of τt to τht is 8:1. The sporadic attenuation of the endothelium, whereby extremely thin areas are generated without risking the mechanical integrity of the lung, appears to be a common scheme evoked in the design of the gas exchangers of the higher vertebrates to maximize transfer of O_2 without compromising the mechanical integrity of the blood-gas (tissue) barrier.

Compared with non-flying mammals, bat lungs have relatively thin blood-gas (tissue) barriers (Maina et al. 1982; Maina and King 1984; Maina et al. 1991). The thinnest barrier (τht) of 0.1204 μm has been found in the flying fox, *Phyllostomus hastatus* (Maina et al. 1991). In the lungs of mammals and birds, the thickness of the blood-gas (tissue) barrier appears to have been optimized (e.g. Maina 1989b; Maina et al. 1989a; Maina 2000b,c). Although mammals span an enormous range of body mass from the minute 2.6-g Etruscan shrew, *Suncus etruscus,* to the about 150-tonne blue whale, a factorial difference of about 60 × 106, the τht in the lung of the shrew of 0.27 μm (e.g. Gehr et al. 1980) differs from that of 0.350 μm in the lung of the whale (Henk and Haldiman 1990) by a factor of only 1.3. In birds the τht of the blood-gas (tissue) barrier in the 7.3-g violet-eared hummingbird, *Colibri coruscans,* is 0.099 μm (Dubach 1981), whereas that of the ostrich, *Struthio camelus* (40 kg), is 0.56 μm (Maina and Nathaniel 2001), a body mass factorial difference of over 5×103, but a difference of the τht of the blood-gas (tissue) barrier of only about 6.

In the highly metabolically active tissues of insects, such as the flight muscle, the tracheoles are never more than 0.2–0.5 μm from a mitochondrion and in some tissues they may be as close as 0.005 μm (e.g. Maina 1989b). The mitochondria normally cluster around the tracheoles (Fig. 68), forming what has been described as a 'mitochondrial continuum'. In the trachea and those tracheoles that lie outside of the tissue cells, O_2 must diffuse through the tracheal wall, the haemolymph and the cytoplasm (the air-etissue barrier) to reach the mitochondria (Schmitz and Perry 1999). In the tick *Dercentor variabilis* and the millipede, *Fontaria* sp., the thickness of the intertaenidial membrane is no more than 0.02 μm (Richards and Anderson 1942). In the stick insect, *Carausius morosus,* the τht of the tracheal epithelium is smallest in the narrowest trachea (Schmitz and Perry 1999); the value increases with the tracheal diameter. In the dragonfly, *Aeschna,* tracheal walls are about 10–30 nm thick (Weis-Fogh 1964a). Although the trachea and the tracheoles are circular in cross section (Figs. 14, 67–72, 74, 76, 77, 102, 103) and from the support provided by the circular and helical taenidial rings can resist compression, depending on need and circumstances, they can readily be shortened or stretched in an accordion manner (Wigglesworth 1974); their volumes can change by as much as 20%–30%.

12 Exposure and Disposition of Respiratory Media in Gas Exchangers

The external gas exchangers that have evolved in higher organisms – fish gills, bird lungs and alveolar lungs of amphibia, reptiles and mammals have some basic features in common, irrespective of the different principles that determine their functioning in detail.

Weibel (1984)

Optimal exposure of the external respiratory medium (water/air) to blood is requisite for efficacious flux of the respiratory gases in a gas exchanger. The presentation of the respiratory media is a dynamic process that is determined by the geometry and the spatial arrangement of the conduits and structures that transport the respiratory media to and within the gas exchanger. Different designs (Fig. 1) have given rise to the different presentations (Fig. 3). Exposition, on the other hand, connotes a steady state in which blood is exposed to an external respiratory medium. In gas exchangers, vascularity and prominence (bulging) of the blood capillaries over the respiratory surface (Figs. 20, 48, 54, 55, 85–87, 110, 112, 115, 116, 120, 122–125, 127–133, 135–137) augment the respiratory surface area while enhancing the exposure of pulmonary capillary blood to the external respiratory medium. In the mammalian lung pulmonary capillary blood is exposed to air essentially on the two sides of the interalveolar septum (Figs. 20, 36, 84–86, 97, 110, 125); such an arrangement of the blood capillaries is called a 'single capillary system'. In the avian lung, about 90% of the exchange the tissue comprises closely interdigitated air and blood capillaries (Maina 1989b; Maina et al. 1989a) (Figs. 6, 7, 8, 17, 57–62, 65); the blood is literally suspended in a three-dimensional space, that is, it is exposed to air virtually on all sides (Figs. 8, 57, 59, 60, 62, 126). The arrangement maximizes the respiratory surface area, promoting and explaining the relatively higher pulmonary diffusing capacity of the avian lung for O_2 (Maina 1989b, 1993, 2000b, c; Maina et al. 1989a). In the African lungfish, *Protopterus aethiopicus* (Maina and Maloiy 1985; Maina 1987a) (Figs. 129, 137), amphibians (Figs. 123, 127, 128) and in certain reptilian lungs (Figs. 120, 122, 130, 132, 136), a double capillary arrangement, that is, where the pulmonary capillary blood is exposed to air only on one side, occurs. In the South American lungfish, *Lepidosiren paradoxa* (Hughes and Weibel 1976), the volume of the pulmonary capillary blood constitutes 3.5% of the total volume of the lung, whereas in mammals (e.g. mouse), the value is 14% and in birds 25% (e.g. Duncker 1974). In the suprabranchial membrane (Figs. 19, 116) and the labyrinthine organ (Figs. 19, 48, 124), the accessory respiratory organs of the catfish *Clarias mossambicus* (Maina and Maloiy 1986), the lung of the slug *Trichotoxon copleyi* (Maina 1989e) (Figs. 28, 133) and the blood vessels on the respiratory groove of the swampworm, *Alma emini* (Maina and Maloiy 1998; Maina et al. 1998) (Fig. 41), only one layer of blood capillaries that protrude into the air space occur. In such a design, strictly speaking, neither the term 'single capillary' nor the term 'double capillary' arrangement applies. Regarding the exposure of blood to an external respiratory medium in a gas exchanger, this is perhaps the most elementary design that has ever evolved. It is, however, pertinent to note that in such designs, rather than the blood in the capillary system being suspended in air/water, it is applied on a supporting (fixed) surface. Hence, the only realizable respiratory construction is one where blood is exposed to a respiratory medium only on one side. In the accessory respiratory organs of the African catfish, *Clarias mossambicus*, to compensate for an inferior design, the thicknesses of the blood-gas (tissue) barrier of the suprabranchial chamber membrane (0.313 μm) and the labyrinthine organ (0.287 μm) are much smaller than that of the water-blood barrier of the gills (17.30 μm) (Maina and Maloiy 1986), a condition that promotes the diffusing capacity of the accessory respiratory organs. The exposure of blood to water in the vascular channels of the secondary lamellae, hemispherical flaps that are located on opposite sides of the filaments of fish gills (Figs. 46, 52, 54, 55, 107, 111) and the gill filaments of crustacean gills (Figs. 47, 49, 66, 121), occurs across two surfaces.

The capillary loading, the ratio between the volume of the pulmonary capillary blood to the respiratory area, is higher in the double capillary than in the single capillary system. A single capillary scheme grants better exposure of blood to air and hence offers a more efficient gas exchange design than a double capillary system. Expectedly, in the vertebrate gas exchangers, the exposure of pulmonary capillary blood to external respiratory media perfects from those of the ectotherms (e.g. fish, lungfishes, amphibians and reptiles) to those of the endotherms (mammals and birds). In mammals, regarding the development and arrangement of the pulmonary blood capillaries of the lung, a recapitulation of phylogeny is manifest. A double capillary arrangement appears during the early stages of the growth of the lung, subsequently converting to a single capillary arrangement from about term (e.g. Pinkerton et al. 1982). In the herbivorous marine mammals, such as the manatees and dugongs, however, a double capillary system persists throughout life (e.g. Henk and Haldiman, 1990). The preponderance of collagen, elastic tissue, smooth muscle and other supporting tissue elements in the interalveolar septa is essential for providing the mechanical support required to counteract the hydrostatic forces during deep dives. A strong fibrous skeleton is necessary to provide support to the lung. In the Weddell seal, *Leptonychotes weddelli*, the alveoli are totally collapsed at depths of 30–35 m, but in that of the dolphin, *Tursiops truncatus*, they do not collapse before a depth of 70 m (e.g. Falke et al. 1985). A double capillary arrangement, that is, a high pulmonary capillary loading, appears to be the price that the cetaceans have had to pay to subsist under water. The lower metabolic rate of the cetaceans may have allowed them to retain an inferior design of their lungs for animals at their phylogenetic level of development.

13 Pneumocytes, Surfactant and Design of Gas Exchangers

The availability of multiple solutions to a given structural problem is one of the reasons why animals have evolved into such an enormous variety of forms.

Gordon (1988)

According to Young-Laplace's law, the pressure (P) needed to distend a spherical body is directly proportional to the surface tension at the air-liquid interface (T) and inversely proportional to the radius of curvature (r^{-1}), i.e. $P = T \quad r^{-1}$. Although an intensely subdivided lung grants an extensive respiratory surface area (Sect. 10), the process engenders higher ventilatory cost (e.g. Wilson and Bachofen 1982). Therefore, for optimal design, a compromise must be conceived between fabrication of an extensive respiratory surface area (by intensification of internal subdivision of a gas exchanger) with minimization of respiratory cost of operation. The surfactant [a mixture of phospholipids (dipalmitoylphosphatidylcholine), neutral lipids and a small quantity of proteins] lines the respiratory surface of the air bladder (e.g. Maina et al. 1996a; Maina 2000d) (Figs. 63, 112) and all those gas exchangers (lungs) that subsequently developed from it.

The process of formation of the surfactant has remained unchanged for at least the last 300 million years (e.g. Power et al. 1999). Lamellar bodies have been described in the lungs of ancient fish such as the lungfishes (Dipnoi) (e.g. Hughes and Weibel 1976; Maina and Maloiy 1985) (Figs. 32), the bichirs, *Polypterus delhezi* and *P. ornatipinnis* (e.g. Zaccone et al. 1989), the gar-fish, *Lepisosteus osseous* and the bowfin, *Amia calva* (e.g. Hughes and Morgan 1973). The surfactant appears to have evolved in the ancestral lungs mainly to protect the epithelial surface (Liem 1987). With the development of more complex lungs, when the buccal force pump could no longer adequately ventilate intensely partitioned lungs, conceivably, the surfactant was assigned the role of attenuating the effects of surface tension at the air-liquid interface of the lung. This was necessary to minimize respiratory work and stabilize the small terminal gas exchange components.

In different forms and quantities, the surfactant is widely distributed in the vertebrate lung (e.g. McGregor et al. 1993). The preponderant O_2 tension (e.g. Acarregui et al. 1995) and physiological factors such as metabolic rate, physical activity, body temperature and ventilation (e.g. Codd et al. 2000) may regulate its composition and function. Hyperventilation and increased sympathetic output, factors indicative of stress, also determine the quantity and the composition of the surfactant (e.g. Orgeig et al. 1995). The biochemistry and molecular bonding of the surfactant is extremely fluid (e.g. Veldhuizen et al. 1998). This allows shifts in the interfacial tension as the terminal air spaces of compliant lungs change in volume. The surfactant is continually cleared and recycled with a turnover of only a few hours. In mammals, the most important stimuli that influence surfactant secretion include stretching of the basement membrane of the type II cells, adrenergic neurotransmitters and temperature (e.g. Codd et al. 2000). Occurring in relatively large quantities (e.g. Vergara and Hughes 1981), the amphibian pulmonary type surfactant may or may not be discharged onto the respiratory space as the tubular myelin type (e.g. Bell and Stark-Vancs 1983). The surfactant of the amphibian lung is not as surface active as that which lines the lungs of the higher vertebrates (e.g. Daniels et al. 1989).

Where exceptionally large terminal air spaces exist, for example, in the simple lungs of the lungfish, amphibians and reptiles (Figs. 15, 21–24, 26, 27, 29, 80–83, 105, 106, 108, 109), the function of the surfactant, as regards reduction of surface tension, is equivocal. The surfactant is, however, known to play various other roles such as prevention of transendothelial transudation of blood plasma across the blood-gas (tissue) barrier, immune suppression, chemotaxis of macrophages, displacement and clearance of deposited particles and has an antioxidant function (e.g. Daniels et al. 1993). In the agamid lizard, *Ctenophorus nuchalis*, the salamander, *Amphiuma tridactylum*, and the siren, *Siren intermedia*, the surfactant-like lipids are envisaged to act as antiglue, that is, they prevent epithelial adhesion after collapse of the lungs, for example, during apnea (e.g. Daniels et al. 1993). The antiglue role of the surfactant may be important in the lungs of those animals that lack well-formed intrapulmonary conducting airways and a diaphragm (e.g. amphibians, reptiles and birds) where the lung is highly susceptible to compression by shifting intracoelomic organs (e.g. Daniels et al. 1994). In the anaconda, *Eunectes murinus*, the surfactant lining occurs both in the anterior faveolated and the posterior saccular parts of the lung (e.g. Phleger et al. 1978). In the faveolated region, the surfactant may play a surface tension-reducing role, whereas in the saccular region it may inhibit extravasation of materials from the pulmonary vasculature onto the surface of the lung. The presence of the surfactant on the air capillaries of the avian lung, gas exchange components that are virtually rigid (e.g. Macklem et al. 1979) (Figs. 7, 8, 57–60, 62, 65), is intriguing. Although it may be simply a phylogenetic carry over from the ancestral reptilian lung (e.g. Perry 1989), in the avian lung the surfactant may well function to curtail transendothelial filtration of fluids from the blood capillaries, averting flooding of the extremely narrow air capillaries. In birds, this may be particularly important because the systolic blood pressure and, by inference, the pulmonary capillary intramural pressure is high. In general, the systolic blood pressure of birds is higher than that of mammals (e.g. Seymour and Blaylock 2000): a systolic pressure of 53 kPa has been reported in the domestic turkey, *Meleagris gallopavo* (Speckman and Ringer 1963). The ventricular wall stress is 38 kPa in the turkey, 15 kPa in the pigeon and 11 kPa in the domestic fowl (Seymour and Blaylock 2000).

The surfactant is conspicuously absent in the gills and in the accessory respiratory organs of the bimodal breathing fish (e.g. Maina and Maloiy 1986). Generally, in reptiles, mammals and birds, the surfactant is secreted by highly differentiated (granular = cuboidal) cells, the type II cells (Figs. 13, 30–36, 64, 88, 96, 120). The type I (squamous = smooth) pneumocytes have extremely thin and extensive cytoplasmic extensions; they cover much of the respiratory surface (e.g. Crapo et al. 1983) (Figs. 86, 95, 97–99, 125, 126, 134). The type II cells have microvilli on their free epithelial surface and are endowed with abundant organelles such as rough endoplasmic reticulum, mitochondria and Golgi bodies (Figs. 13, 30, 31, 34, 35, 64, 88, 120, 136). Compared with the type I cells, the highly attenuated cytoplasmic extensions of which are devoid of organelles, the type II cells are evidently relatively metabolically more active. The lamellated bodies are the precursors of the surfactant (Figs. 31–36, 63, 88, 96, 120, 123) . They are discharged onto the respiratory surface by merocrine secretion (Fig. 13). In the amphibian lungs (Figs. 31, 127) and lungfishes (Dipnoi) (Fig. 32, 129), the pneumocytes are generally undifferentiated (e.g. Hughes and Weibel 1976; Maina 1987a; Goniakowska-Witalinska 1995). The pneumocytes in the more primitive reptiles are undifferentiated, whereas those in the relatively more advanced ones are differentiated (e.g. Perry et al. 1989) (Figs. 35, 136). In mammals (Figs. 13, 30, 33, 36, 88) and birds (Fig. 96), the pneumocytes are completely differentiated into types I and II cells (e.g. Maina 1987b; Daniels et al. 1990).

The differentiation of the pulmonary pneumocytes in the higher vertebrates may be adaptive. In the human lung, the extremely thin metabolically inactive type I cells cover the greatest extent of the respiratory surface (Crapo et al. 1983). This produces a thin blood-gas (tissue) barrier. In the air breathing vertebrates, except in the primitive lungs, such as those of the caecilian *Boulengerula taitanus* (Maina and Maloiy 1988) (Fig. 123) and the lungfish *Protopterus aethiopicus* (Maina and Maloiy 1985; Maina 1987a) (Fig. 129), where the perikarya of the undifferentiated pneumocytes are located over the blood-gas (tissue) barrier, the type II cells are normally located in crevices between the blood capillaries (Figs. 34, 88, 120, 127, 130, 135), that is, away from the blood-gas (tissue) barrier. An extreme case is found in the avian lung, in which the perikarya of the type II cells are entirely situated in the atria and infundibula of the parabronchi (Fig. 64), that is, away from the gas exchange site (Figs. 6, 17, 18, 114). The perikarya of the type I cells are seldom found in the exchange tissue (Fig. 62). The location of the cell bodies away from the respiratory site accounts for the particularly thin blood-gas (tissue) barrier in the avian lung (Maina 1989b, 2000b, c; Maina et al. 1989a). The location of the type II cells in the corners of the terminal air spaces where they are most sensitive to the mechanical distorsions and stretching of the lung may initiate surfactant secretion in response to changes in breathing patterns (e.g. Wirtz and Dobbs 1990).

14 Fractal Geometry and Design of Gas Exchangers

Although the materials found in biology are often very different from those used in engineering, the geometries of the structures in which materials can be employed to carry loads are generally much the same. Nature is frequently more clever than engineers at developing the potential of a given structural concept.

Gordon (1988)

The airways of the mammalian lung form a complex multigenerational dichotomous branching arrangement (Figs. 138, 141, 142). The trachea corresponds to the stem of a botanical tree and the alveoli, the terminal gas exchange components (Figs. 78, 79, 84, 85), to its leaves. The arterial and venous systems of the pulmonary vascular system generally follow congruent bifurcative patterning (Figs. 139–142). Since tree-like designs are produced by external and internal factors engaging the genotype, they should manifest optimal morphological constructions. On tree-like designs, McMahon and Kronauer (1976) observed that 'every tree is continually sensing its own overall geometry, altering its proportions in such a way as to keep that geometry stationary during growth'. In the physics of internalized flow, the size, shape and geometry of the confining space regulate the flow of the transmitted fluid. In the gas exchangers, the morphoarchitectonics of the conducting systems should set the flow dynamics of the respiratory fluid media, namely, water, air and blood. It is conceivable that over the evolutionary continuum, the conducting systems have been refined to grant cost-effective transfer and optimal presentation and exposition of the respiratory media (Sect. 12). Regarding aspects such as minimization of resistance, work and power loss, from engineering and mathematical fluid flow analyses, it has indeed been shown that the lung has contracted premium designs. Woldenberg et al. (1970) termed the prevalent states in which the erected designs have conferred efficient flux of air and blood simply 'law and order'.

Models for analysing and quantifying tree-like structures were first developed by geomorphologists to define river drainage patterns. The models were subsequently adopted and in certain cases modified for analysis of biological structures, including the tracheobronchial and vascular systems of the lung (e.g. Weibel, 1963). The main ones are: (a) the 'inverted-tree model', in which the dimensions of the branches are measured from a stem branch to a set point at the periphery (e.g. Weibel 1963), (b) the 'order-based (retrograde = converging) model', in which the branches (orders) are measured from a peripheral point towards a parent order (e.g. Singal et al. 1973) and (c) the 'diameter-based statistical reconstruction model', in which parts of a conducting system are categorized without regard to their actual topographical locations (e.g. Bates 1993). From its simplicity and intuitive appeal, in that the analysis is made along the flow (functional) trajectory as well as down the developmental axis, compared with other models, Weibel's Model A (Weibel 1963), the 'regular (= symmetrical) dichotomy model', has been more widely used to quantify the architecture of the airway and vasculature of the mammalian lung. The model allows the structural association of the various generations to be readily constructed and the fluid flow dynamics to be calculated after only elementary assumptions.

Although simple bifurcation models grant reasonable conceptual approximations to the organization of the airway and vascular systems of the lung, they unfortunately idealize them by oversimplifying the real geometry. By discounting important structural features for the sake of simplicity, such models fail to adequately explain the mechanisms that are caused by, occur in or are dependent on asymmetrical patterning. In the human lung, for example, Nikiforov and Schlesinger (1985) found remarkable inter- and intrasubject differences in the bronchial geometry, with asymmetrical bifurcation prevailing. Bates (1993) observed that the airways of a given generation are not identical but rather vary in their wall thicknesses, quantity of smooth muscle and number of parenchymal attachments. Among the bronchial systems of human, rat, mouse and monkey lungs, Patra (1988) noted significant intra- and intersubject differences in the bifurcation patterns. Regarding the total number of airway generations in different regions of the human lung, Weibel and Gomez (1962) observed marked variations. Nikiforov and Schlesinger (1985) noted greater variability between length and bifurcation than between diameter and bifurcation. According to Hammersley and Olson (1992), asymmetrical bifurcation of the airway and vascular systems of the lung may allow more effective three-dimensional space filling, may be the most appropriate design compatible with normal development of the lung and may grant consequential shifts in the air and blood flow through the gas exchanger. The essence of the extravagant bifurcation and extreme folding and compression of the bronchial and vascular systems of the lung (Figs. 5, 6, 8, 15, 17–19, 27, 28, 57, 65, 78, 79, 105, 109, 135, 138–142) is fundamental to a vast respiratory surface area in a confined space (Sect. 10). The total number of branches at a given generation (n) of a dichotomously branching tree-like designed structure, as in the tracheobronchial and vascular systems of the mammalian lung, is estimated as $2n$ (e.g. Weibel and Gomez 1962; Weibel 1963, 1984). In the 23 generations of the bronchial system measured in the lung of the domestic pig, *Sus scrofa,* by Maina and Nathaniel (2001), over 8 million branches of a diameter above 0.5 mm were estimated.

Until fairly recently, there were no satisfactory mathematical methods of quantifying and defining the morphological organization of complex biological structures. The application of conventional morphometric methods to determine parameters such as size and shape was based on the erroneous assumption that the structures approximated regular geometrical objects. Rather than accurately quantifying intricate form, the classical Pythagorean (6 b.c.) and Euclidean (300 b.c.) geometries that use straight lines and smooth and regular curves to make flat shapes and figures such as squares, triangles and circles and in which space is defined in terms of discrete dimensions, that is, a point has no (zero) dimension, a line has 1 dimension, a plane (area) has two and a solid (volume) three dimensions, idealize it. The conventional integer (quantum = discontinuous) dimensions are unsatisfactory in describing the topology of highly complex biological structures that lack specific scales of size. A fractal dimension strictly defines the extent to which the geometrical properties of a complex structure fit between the integer (Euclidean) dimensions. Fractal geometry has thus come to be termed the 'geometry between dimensions' (e.g. Briggs 1992). A folded structure is topologically in transition from a smooth, flat, two-dimensional surface to a three-dimensional volume. Contingent upon the degree and the extent of folding, highly corrugated surfaces have fractal dimensions that range between 2 and 3. The surface fractal dimension of the peripheral airways of the mammalian lung is about 3 (e.g. Mandelbrot 1983) or 2.5 (e.g. Weibel 1963). These high values show that the airways have branched so intensely as to virtually fill a three-dimensional space. The pulmonary vasculature itself is so greatly folded that it has an effective fractal dimension of 3, with the arteries alone having that of 2.7 (Briggs 1992).

Manifesting the importance of respiration for life, most gas exchangers are fabricated by molding three fractal entities. These are the bronchial (Fig. 138), the arterial (Fig. 139) and the venous (Fig. 140) systems (e.g. Weibel 1986). In the human lung, the fractal dimensions of the diameter elements of the arterial and the venous tree are 2.71 and 2.64 respectively, whereas the equivalent values for the length element are 2.97 and 2.86, respectively (Huang et al. 1996). In the dog lung, the fractal dimension of the blood flow is 1.22 (Barman et al. 1996). Pennycuick (1992) suggested that the morphological complexity of the avian pulmonary system has resulted in a high fractal dimension of about 2.5, allowing birds to attain exceptional respiratory capacities that support locomotory activities such as sustained flights at high altitude. The highly efficient tracheal system of insects is thought to have a high fractal surface dimension, whereas fish gills have a value of 2 (Pennycuick 1992). The fractal properties of the airways and the microvascular network of the lung are envisaged to allow an extensive internal surface to be ventilated and perfused at low energy cost (e.g. Weibel 1991, 1996).

Discovered and advanced between the early 1960s and 1970s by Benoit Mandelbrot (Mandelbrot 1983), fractal geometry provides a powerful method for characterizing the form and design of multiple scaled (irregular = complex) structures. The word 'fractal' implies 'fractured' and 'fractional', that is, a geometry that focuses on disconnected, folded and rough shapes that can only be adequately described by between integer, that is, fragmented, dimensions. The great importance of fractal geometry in biological morphometry is expressed in the following statements/observations made by certain leading authorities on the subject: Weibel (1984, 1991) and West (1990) pointed out that fractal design allows biological systems to function over wide range of states and conditions without failure; West (1987) remarked that the fluidity intrinsic to a fractal design might be important in directing and supporting the trial and error process of evolution; Pennycuick (1992) speculated that application of fractal geometry in biological design should allow structural schemes to scale over a wide range of body mass without necessitating drastic changes in the basic morphology; West et al. (1999) simply termed fractal geometry the 'fourth dimension of life' and Weibel (1984) asserted that 'the fractal property of the airways and blood vessels in the lung probably offers one of the most persuasive cases upholding the view that biological design endeavours for optimal bioengineering states'. Among the Etruscan shrew, *Suncus etruscus*, the smallest and the most highly metabolically active extant mammal, the African elephant, *Loxodonta africana*, the largest terrestrial mammal, and the blue whale, the largest extant mammal, animals that differ enormously in body size, the lungs are fundamentally morphologically similar (e.g. Gehr et al. 1980; Henk and Haldiman 1990). The structure of the lung of the minute 7-g violet-eared hummingbird, *Colibri coruscans* (e.g. Dubach 1981), is fundamentally similar to that of the huge ostrich, *Struthio camelus* (Maina and Nathaniel 2001). Although an application that may potentially become an inclusive biological principium for analysis and understanding of physical and biological design, presently, fractal geometry largely remains unknown, unappreciated and underutilized by the vast majority of researchers.

15 Sheet Flow Design in the Vasculature of Gas Exchangers

One can visualise the production of the final integrated morphological structure as a complicated developmental choreography in which the ontogenies of the separate morphogenetic components are spatially and temporally co-ordinated.

Atchley and Hall (1991)

The geometric arrangement and organization of the blood capillary systems of the gas exchangers is remarkably different from those that appertain to the systemic circulation. Whereas, for example, the blood capillaries in the muscle tissue occur as long, loosely connected structures (e.g. Mathieu-Costello et al. 1992), the pulmonary blood capillaries form an extremely dense network. Rather than passing through a system of long tubes, the pattern of flow of blood in a classic capillary system, in the gas exchangers, the blood forms an expansive film, a 'sheet'. Various investigators have analysed and mathematically modelled the size, shape and geometry of the blood capillaries of various gas exchangers to gain an insight into the blood flow dynamics and the rate of gas transfer at the water/air-blood (tissue) interface. The observations and inferences that have been made conflict even on certain fundamental aspects. For example, Guntheroth et al. (1982) regarded the mammalian pulmonary blood capillaries as 'intersecting tubules', whereas Weibel (1963) and Weibel and Gomez (1962) considered them to be 'intersecting, short, circular, cylindrical tubules arranged predominantly in hexagonal arrays'. Hijiya and Okada (1978) suggested that the arrangement of the alveolar blood capillaries is 'more frequently pentagonal in shape'. Schraufnagel et al. (1986) considered the shape of the alveolar blood capillaries to change from 'ring' to 'square' configurations between the inspiratory and expiratory phases. On the basis of what they considered to be 'short and closely knit blood capillaries where the capillary segments were wider than they were long' and the haemodynamic properties where the vascular compliance was such that 'only sheet thickness increased when intravascular pressure was raised', Fung and Sobin (1972) and Sobin et al. (1979) formulated what they named a 'sheet-flow model' to analyse the alveolar blood volume and transit time. They considered the pulmonary capillary system to comprise of a construction made up of cellular/tissue 'posts' that traversed a vascular space.

Morphologically, the respiratory microcirculatory systems of gas exchangers comprise complex vascular networks (Figs. 7, 8, 20, 48, 52–58, 60–62, 79, 84–87, 104, 107, 111, 112, 114, 115, 117, 134–136, 139–142). Indeed, microvascular surface density is more or less an indicator of a respiratory surface and its actual intensity is a relative mark of functional efficiency. The size and geometry of a capillary system determine the rheology of the blood flow and hence the transit time of blood across a sector of blood capillaries. Weibel and Gomez (1962) estimated the number of capillary segments in the human lung to be 277 billion! As pointed out in Sect. 12, optimal exposure of blood to air/water is a necessary feature for efficient gas exchange. In the human lung, about 213 cm^3 of the pulmonary capillary blood is spread over a respiratory surface area of about 150 m^2 and the air and blood are separated by a tissue barrier that is only 0.65 μm thick (Gehr et al. 1978). In simpler terms, 1.5 cm^3 of blood (about 18 drops) are spread over an area of 11/2 m^2 and the gas exchange media are separated by a tissue barrier that is 50 times thinner than that of a postage stamp! Calculated as a ratio between the pulmonary capillary blood volume and the surface area of the blood-gas (tissue) barrier, in an adult person, the film of blood that overlies the respiratory surface is on average only about 1.4×10^{-6} μm thin. In the violet-eared hummingbird, *Colibri coruscans*, in which the volume of the pulmonary capillary blood is 0.046 cm^3 and the surface area of the blood-gas (tissue) barrier is 0.054 m^2 (Dubach 1981), the film is 9×10^{-8} μm thin! At term, the human placenta contains 45 cm^3 of capillary blood that is spread over an area of about 11 m2 (Aherne and Dunnill 1966).

Except in the tracheates, where O_2 is delivered directly to the tissue cells from the atmosphere by diffusion through contiguous conduits, the trachea (e.g. Maina 1989a) (Figs. 67–72, 102, 103), with certain structural variations and modifications, sheet-flow design has developed in virtually all the other kinds of gas exchangers (e.g. Farrell et al. 1980; Maina 2000a). As a general definition, the design consists of two diametrically located thin, flat, parallel 'bands' of epithelial cells and connective tissue elements that are joined by cellular or connective tissue components that in a 'pillar' or 'column'-like manner traverse a vascular space. The lung of the slug, *Trichotoxon copleyi* (Maina 1989e) (Figs. 28, 133), fish gills (Maina, 1990a) (Figs. 52–56, 107, 111), crustacean gills (Maina 1990b) (Figs. 66, 121), amphibian lungs (Maina and Maloiy 1988; Maina 1989c) (Fig. 127), reptilian lungs (Maina 1989d; Maina et al. 1989b) (Figs. 104, 119, 120) and mammalian lungs (e.g. Maina et al. 1982; Maina 1988) (Figs. 20, 87, 88, 111) display a sheet-flow design: connective tissue elements (Figs. 123, 128) and endothelial cells (Figs. 53–56, 66, 107, 119, 120, 122) form 'struts' or 'columns' that couple the diametrically located epithelial cells. In the avian lung, in which the blood (Figs. 59–61) and air (Figs. 65, 113) capillaries interdigitate profusely in three dimensions (Figs. 6–8, 57, 58, 62, 126), the sheet-flow design, in which the exposure of blood to water/air is essentially confined to two opposite surfaces (Figs. 20, 47, 49, 50, 53, 55, 107, 111, 121), does not occur. The pulmonary capillary blood is virtually suspended in three-dimensional space (Figs. 57–60, 62, 126), affording excellent exposure of blood to air. Air capillary to air capillary and blood capillary to blood capillary contacts (Figs. 58, 62, 126) occur only rarely and, when they do occur, only over minimal distances.

In the gas exchangers, the sheet-flow concept offers an empirical explanation of the basis of collateral perfusion of adjacent microvascular units if and when the most direct route of blood supply is interrupted. The pathways that remain open are essentially channels of the same broad microvascular sheet that is supplied with blood by a pulmonary artery and drained by a pulmonary vein. The network guarantees that gas exchange continues even after vascular collapse or obstruction occurs in some parts of a respiratory organ. Owing to the physical interruptions caused by the 'posts' or 'columns', the blood flow in the microvascular maze of narrow channels is circuitous rather than direct. This slows down the blood flow and may generate some turbulence, features that may promote gas exchange. At rest, when the O_2 need is low, a substantial portion of the blood capillaries is unperfused or underperfused (e.g. Okada et al. 1992), but during activity, that is, when O_2 consumption increases, the collapsed channels are recruited. In the mammalian lung, a pulmonary capillary network may continue outside of a pulmonary acinus (e.g. Weibel 1984), a respiratory unit that consists of some 10^4 alveoli (e.g. Haefeli-Bleuer and Weibel 1988). In the human lung, about 50% of the total volume of blood in the lung is found in the interalveolar septa (Burri 1985). The intrapulmonary capillaries constitute more than 90% of the capillary bed (König et al. 1993).

Pulmonary circulation is fundamentally a low-pressure, low-resistance system (e.g. West 1974). In the human being, the blood pressure in the pulmonary artery is 2 kPa compared with 13 kPa in the systemic circulation. The resistance to blood flow across the pulmonary blood capillaries (diameter 10–14 μm) is so low that 5–10 l of blood can pass through the lung each minute under a pressure of less than 1.3 kPa. Blood flow across the lung is determined by the pressure differential between the arterial and the venous ends and the resistance in the intercalated capillary bed. Much of the pressure drop in the pulmonary circuit occurs in the blood capillaries. Low pulmonary capillary transmural blood pressure warrants the structural integrity of the capillary network and minimizes exudation of blood plasma onto the respiratory surface. Okada et al. (1992) found that resistances vary significantly between different capillary segments in a single alveolar wall, and Schraufnagel (1987) observed that pulmonary blood capillaries are uniform in size and have no directionality. The vascular channels are not simply coupled in series or in parallel but are made up of irregular arcades and shunt pathways that regulate the mechanism of blood flow and its distribution. Pulmonary microvascular pressure is pulsatile (Maarek and Chang 1991); spontaneous cyclic contractions of the blood capillaries have been reported. The volume of blood in the pulmonary circulation determines various physiological processes, including the diffusing capacities of the gas exchangers (e.g. Roughton and Forster 1957).

Unlike in mammals, in which the lung has its own circulatory arch that lies in series with the systemic one, fish gills are located on the arterial side of the circulation. The transmural pressure across the microvascular units of the secondary lamellae may be as high as 12 kPa (e.g. Hughes 1976) compared with that of only about 1.1 kPa in the pulmonary capillary system of the human lung (West 1974). The greater thickness of the water-blood barrier in fish gills (e.g. Hughes and Morgan 1973; Maina et al. 1996b) compared with the blood-gas (tissue) barrier in mammalian and avian lungs (e.g. Gehr et al. 1981; Maina et al. 1989a) can be ascribed to the higher transmural blood pressure in the lamellar microvasculature and the presence of a denser external supporting fluid medium (water).

16 Summary and Conclusion

The task of designing or developing an efficient structure is not an easy one. In biology, millions of years of natural selection have provided an encyclopedic variety of imperfect or intermediate structures.

Gordon (1988)

Over its life, an organism's survival and success are determined by the inventory of vital adaptations that its progenitors have creatively appropriated, devised and harnessed along the evolutionary pathway. Such conserved attributes provide the armamentarium necessary for withstanding the adverse effects of natural selection. Refinements of the designs of the respiratory organs have been critical for survival and phylogenetic advancement of animal life. Gas exchangers have changed in direct response to the respiratory needs of whole organisms in different environmental states and conditions. Nowhere else is the dictum that in biology 'there are no rules but only necessities' more manifest than in the evolutionary biology of the gas exchangers. The constructions have been continually fashioned and refined to meet specific needs. Solutions to common respiratory needs have been typified by profound structural convergence. Over the evolutionary continuum, as shifts in environmental situations occurred, infinitely many designs should theoretically have emerged. Moreover, without specific selective pressures and preference for certain designs, considering that there are only two naturally occurring respirable fluid media (air and water), air-lungs, water-lungs, air-gills and water-gills would have formed to similar extents. Factors such as body size, phylogenetic level of development, respiratory medium utilized and habitats occupied have permutatively prescribed the design of the gas exchangers. The construction of the modern gas exchangers has eventuated through painstaking cost-benefit analysis. Trade-offs and compromises have decreed only a limited number of structurally feasible and functionally competent outcomes. The morphological congruity (analogy) of the gas exchangers indicates that similar selective pressures have compelled the designs. Solutions to metabolic demands for molecular O_2 have only differed in details. Passive physical diffusion, for example, is the ubiquitous method of transfer of O_2 across biological tissues. Gills, evaginated gas exchangers, were the primordial respiratory organs that evolved for water breathing, whereas lungs (invaginated gas exchangers) developed for terrestrial (air) breathing. Transitional (= bimodal = amphibious) breathing has evolved in animals with specialized organs that extract O_2 from both water and air. Lungs are tidally (= bidirectionally) ventilated, while gills are unidirectionally ventilated, a feature that allows the highly efficient counter-current disposition between blood and water. Since animals occupy inconstant environmental milieus and their metabolic states vary, gas exchangers are designed to operate optimally across a spectrum of conditions that range from resting to exercise and even under hypoxia. Inbuilt structural and functional flexibility provides the requisite safety factors that allow adjustments to modest pressures.

The fundamental structural features that determine the respiratory function of a gas exchanger are respiratory surface area, thickness of the blood-water/gas (tissue) barrier and volume of the pulmonary capillary blood. The diffusing capacity of a gas exchanger correlates directly with the surface area and inversely with the thickness of the blood-water/gas (tissue) barrier. An extensive surface area is generated in gills by extensive stratification of the gas exchanger and in lungs by profuse internal subdivision. Compartmentalization yields small terminal gas exchange compartments that compel greater commitment of energy to ventilate. The surfactant, a phospholipid lining, reduces the forces of surface tension at the air-water interface. This attenuates the propensity of physical collapse of the minute gas exchange units and minimizes the cost of ventilation. The surfactant characterizes all the gas exchangers derived from the piscine air bladder. In the lower air-breathing vertebrates, such as the lungfishes (Dipnoi), amphibians and certain reptiles, the pneumocytes are not differentiated into type I and II cells, as is the case in the lungs of the higher vertebrates-birds and mammals. It is envisaged that in endotherms, the overall numerical density of the pneumocytes and hence the O_2 consumption of the gas exchangers may be reduced and a thin blood-gas (tissue) barrier generated, factors that enhance respiratory efficiency. The thin blood-gas (tissue) barriers, for example, those of the mammalian (in the respiratory sections of the interalveolar septum) and avian lungs, consist of an epithelial cell and an endothelial cell with a common basement membrane. An interstitial space occurs in the blood-air/water (tissue) barriers of the gas exchangers of fish gills and lungs of lungfishes, amphibians, reptiles and in the supportive parts of the interalveolar septum of the mammalian lung. Collagen, elastic tissue, nerves, lymphatic vessels and smooth muscle elements are found in the interstitial space. The thickness of the blood-air/water (tissue) barrier allometrically changes very little. This suggests that the thicknesses of the blood-water/air (tissue) barriers have been optimized. The presentation and exposure to the gas exchange media (water/air to blood), features dictated by the geometry and arrangements of the structural components of the gas exchangers, contribute greatly to respiratory efficiency. The countercurrent presentation between water and blood in fish gills is the most efficient design in the evolved gas exchangers: It was imperative for survival in water, a medium that contains relatively less O_2 and is more expensive to breathe. In the evolved vertebrate gas exchangers, the exposure of blood to air is best manifested in the diffuse design of the avian lung, where the capillary blood is literally suspended in a three-dimensional air space, the blood being exposed to air virtually across the entire blood-gas (tissue) barrier. A double capillary design occurs in the lungs of amphibians and generally those of reptiles, whereas a single capillary design commonly occurs in those of adult mammals. The capillary loading (the ratio of the volume of the capillary blood to the surface area across which blood is exposed to air) in lungs with a double capillary arrangement is high and manifests a poor design. On the other hand, the low capillary loading that characterizes the single capillary system indicates better exposure of blood to air and greater respiratory capacity. Fractal geometry features in the construction of the gas exchangers. The highly versatile design allows the gas exchangers to function optimally under different conditions and circumstances and to maintain congruent morphologies over a wide range of body size, shape and metabolic capacities. At the gas exchange level, sheet-flow design preponderates in the evolved gas exchangers; blood is efficiently exposed to the external respiratory medium. The respiratory capacity of a gas exchanger is comprehensively granted by refinements of structural features and functional processes. Modelling, mathematical integration of structural and functional parameters, provides a holistic view of the essence of the design of a gas exchanger.

References

Abdalla MA, Maina JN, King AS, King DZ, Henry J (1982) Morphometrics of the avian lung. 1. The domestic fowl (*Gallus gallus* variant *domesticus*). Respir Physiol 7: 267–278

Acarregui MJ, Brown JJ, Mallampalli RK (1995) Oxygen modulates surfactant protein mRNA expression and phospholipid production in human foetal lung in vitro. Am J Physiol 268: L818–L825

Aherne W, Dunnill MS (1966) Morphometry of the human placenta. Br Med Bull 22: 5–8

Anderson AE, Felbeck H, Childress JJ (1990) Aerobic metabolism is maintained in animal tissue during rapid sulfide oxidation in the symbiont-containing clam *Solemya reidi*. J Exp Zool 256: 130–134

Atchley WR, Hall BK (1991) A model for development and evolution of complex morphological structures. Biol Rev 66: 101–157

Bakhle YS (1975) Pharmacokinetic function of the lung. In: Junod AF, Haller R (eds) Lung metabolism. Academic Press, New York, pp 293–299

Ball P (2001) Life's lessons in design. Nature 409: 413–416

Barman SA, McCloud LL, Catravas JD, Ehrhart IC (1996) Measurement of pulmonary blood flow by fractal analysis of flow heterogeneity in isolated canine lungs. J Appl Physiol 81: 2039–2045

Bates JHT (1993) Stochastic model of the pulmonary airway tree and its implications for bronchial responsiveness. J Appl Physiol 75: 2493–2499

Bech C, Johansen K (1980) Ventilation and gas exchange in the mute swan, *Cygnus olor*. Respir Physiol 39: 285–295

Bell PB, Stark-Vancs VI (1983) SEM study of the microarchitecture of the lung of the giant salamander *Amphiuma tridactylum*. Scan Electr Microsc 1983: 449–456

Bennett AF (1987) The accomplishments of ecological physiology. In: Feder ME, Bennett AF, Burggrenn WW, Huey RB (eds) New direction in ecological physiology. Cambridge University Press. Cambridge, pp 1–8

Bernhard JM, Buck KR, Farmer MA, Samuel SB (2000) The Santa Barbara Basin is a symbiosis oasis. Nature 403: 77–80

Birks EK, Mathieu-Costello O, Fu Z, Tyler WS, West JB (1997) Very high pressures are required to cause stress failure of pulmonary capillaries in thoroughbred racehorses. J Appl Physiol 82: 1584–1592

Brakenbury JH (1984) Physiological responses of birds to flight and running. Biol Rev 59: 559–575

Briggs J (1992) Fractals: the patterns of chaos. Simon and Schuster, New York

Budgett JS (1900) Observations on *Polypterus* and *Protopterus*. Proc Cambr Philos Soc 10: 236–240

Burri PH (1985) Morphology and respiratory function of the alveolar unit. Int Arch Allergy Appl Immunol 76: 2–12

Callahan PS (1972) The evolution of insects. Holliday House, New York

Carlson CW (1960) Aortic rupture. Turkey Producer, January Issue

Carroll RL (1988) Vertebrate palaeontology and evolution. WH Freeman and Co., New York

Chaetum EP (1934) Limnological investigations on respiration, annual migratory cycle and other related phenomena in fresh water pulmonate snails. Trans Am Microsc Soc 53: 348–407

Clark A, Clark PAA, Connett RJ, Gayaski TEJ, Honig CR (1987) How large is the drop in Po_2 between the cytosol and mitochondrion? Am J Physiol 252: C583–C587

Codd, JR, Slocombe NC, Daniels CB, Wood PG, Orgeig S (2000) Periodic fluctuations in the pulmonary surfactant system in Gould's wattled bat (*Chalinolobus gouldii*). Physiol Biochem Zool 73: 605–612

Corral JLD (1995) Anatomy and histology of the lung and air sacs of birds. In: Pastor LM (ed) Histology, ultrastructure and immunohistochemistry of the respiratory organs in non mammalian vertebrates. Universidad de Murcia Press, Murcia, Spain, pp 179–233
Costello ML, Mathieu-Costello O, West JB (1992) Stress failure of alveolar epithelial cells studied by scanning electron microscopy. Am Rev Respir Dis 145: 1446–1455
Crapo JD, Young SL, Kram EK, Pinkerton KE, Barry BE, Crapo RO (1983) Morphometric characteristics of cells in the alveolar region of the mammalian lung. Am Rev Respir Dis 128: S42–S46
Daniels CB, Barr HA, Nicholas TE (1989) A comparison of the surfactant associated lipids derived from reptilian and mammalian lungs. Respir Physiol 75: 335–348
Daniels CB, Barr HA, Power JHT, Nicholas TE (1990) Body temperature alters the lipid composition of pulmonary surfactant in the lizard *Ctenophorus nuchalis*. Exp Lung Res 16: 435–449
Daniels CB, Eskandari-Marandi BD, Nicholas TE (1993) The role of surfactant in the static lung mechanics of the lizard *Ctenophorus nuchalis*. Respir Physiol 94: 11–23
Daniels CB, Orgeig S, Wilsen J, Nicholas TE (1994) Pulmonary type surfactants in the lungs of terrestrial and aquatic amphibians. Respir Physiol 95: 249–258
de Beer G (1954) *Archeopteryx lithographica*. Br Mus Nat Hist, London
Dejours P (1975) Principles of comparative respiratory physiology, 1st ed. Elsevier-North Holland, Amsterdam
Dejours P (1988) Respiration in water and air: adaptations, regulation and evolution. Elsevier, Amsterdam
Dejours P, Garey WF, Rahn H (1970) Comparison of ventilatory and circulatory flow rates between animals in various physiological conditions. Respir Physiol 9: 108–117
Diamond JM (1998) Evolution of biological safety factors: a cost/benefit analysis. In: Weibel ER, Taylor CR, Bolis L (eds) Principles of animal design: the optimization and symmorphosis debate. Cambridge University Press, Cambridge, pp 21–27
Dubach M (1981) Quantitative analysis of the respiratory system of the house sparrow, budgerigar, and violet-eared hummingbird. Respir Physiol 46: 43–60
Dullemeijer P (1974) Concepts and approaches in animal morphology. Van Gorcum, Assen, The Netherlands
Duncker HR (1971) The lung-air sac system of birds. A contribution to the functional anatomy of the respiratory apparatus. *Ergeb Anat Entwickl* 45: 1–171
Duncker HR (1974) Structure of the avian respiratory tract. Respir Physiol 22: 1–34
Falke KJ, Hill RD, Schneider RC, Guppy M, Liggins GC et al. (1985) Seal lungs collapse during free diving: evidence from arterial nitrogen tensions. Science 229: 556–567
Farrell AP, Sobin SS, Randall DJ, Crosby S (1980) Sheet blood flow in the secondary lamellae of teleost gills. Am J Physiol 239: R428–R436
Fishman AP (1979) Introduction. In: Wood SC, Lenfact C (ed) Evolution of respiratory processes: a comparative approach. Marcel Dekker, New York, pp xi–xiv
Forster RE (1973) Some principles governing maternal-foetal transfer in the placenta. In: Comline KS, Cross KW, Dawes GS, Nathanielz PW (eds) Foetal and neonatal physiology, Proceedings of the Sir Henry Barcroft Centenary Symposium. Cambridge University Press, Cambridge, pp 223–237
French MJ (1988) Invention and evolution design in nature and engineering. Cambridge University Press, Cambridge
Fung YC, Sobin SS (1972) Elasticity of the pulmonary alveolar sheet. Circ Res 30: 451–469
Gardiner BG (1980) Tetrapod ancestry: a reappraisal. In: Panchen AL (ed) The terrestrial environment and the origin of land vertebrates. Academic Press, London, pp 135–207
Gardner MN, Smits AW, Smartresk NJ (2000) The ventilatory responses of the caecilian *Typhlonectes natans* to hypoxia and hyperoxia. Physiol Zool 73: 23–29
Gehr P, Bachofen M, Weibel ER (1978) The normal human lung: ultrastructure and morphometric estimation of diffusion capacity. Respir Physiol 32: 121–140
Gehr P, Mwangi DK, Amman A, Maloiy GMO, Taylor CR, Weibel ER (1981) Design of the mammalian respiratory system: V. Scaling morphometric diffusing capacity to body mass: wild and domestic animals. Respir Physiol 44: 61–86
Gehr P, Sehovic S, Burri PH, Claasen H, Weibel ER (1980) The lung of shrews: morphometric estimation of diffusion capacity. Respir Physiol 44: 61–86

Gil-Vernet JM (1968) Morphology and function of vesico-prostato-urethral musculature. Edzioni Casova, Treviso
Goniakowska-Witalinska L (1995) The histology and ultrastructure of the amphibian lung. In: Pastor LM (ed) Histology, ultrastructure and immunohistochemistry of the respiratory organs in non-mammalian vertebrates. Publicaciones Universidad de Murcia, Murcia, Spain, pp 77–112
Gordon JE (1988) The science of structures and materials. Scientific American Library, New York
Graham JB, Dudley R, Agullar NM, Gans C (1995) Implications of the late Paleozoic oxygen pulse of physiology and evolution. Nature, Lond 375: 117–120
Graiger E, Steinlechner P, Maran R, Méndez G, Eberl T, Margreiter R (1995) Control of mitochondrial and cellular respiration by oxygen. J Bioenerg Biomembr 27: 583–596
Guntheroth WG, Luchtel DL, Kawabori I (1982) Pulmonary microcirculation: tubules rather than sheet and post. J Appl Physiol 153: 510–515
Gyles NR (1989) Poultry, people and progress. Poultry Sci 68: 1–8
Haefeli-Bleuer B, Weibel ER (1988) Morphometry of the human pulmonary acinus. Anat Rec 220: 401–414
Haldane JS (1922) Respiration. Yale University Press, New Haven
Hammersley JR, Olson DE (1992) Physical models of the smaller pulmonary airways. J Appl Physiol 72: 2402–2414
Hanson D, Johansen K (1970) Relationship of gill ventilation and perfusion in dogfish, *Squalus suckleyi*. J Fish Res Bd Can 27: 551–564
Harold FM (1986) The vital force: a study of bioenergetics. W.H. Freeman & Co., New York
Harrison JF, Lighton JRB (1998) Oxygen-sensitive flight metabolism in the dragon fly *Erythemis simplicicollis*. J Exp Biol 201: 1739–1744
Harrison JF, Scott C, Marden JH, Kirkton SD, Rozo A, Yang X (2001) Mite not make it home: tracheal mites reduce the safety margin for oxygen delivery of flying honeybees. J Exp Biol 204: 805–814
Hartwell LH, Hopfield JJ, Leiber S, Murray AW (1999) From molecular to modular cell biology. Nature 40: 47–52
Hellems HK, Haynes FW, Dexter L (1949) Pulmonary capillary pressure in man. J Appl Physiol 2: 24–52
Henk WG, Haldman JT (1990) Microanatomy of the lung of the bowhead whale *Balaena mysticetus*. Anat Rec 226: 187–197
Hijiya K, Okada Y (1978) Scanning electron microscopy of the pulmonary capillary vessels in the rat. J Electron Microsc 1: 49–53
Hill EP, Power GG, Longo LD (1973) Mathematical simulation of pulmonary O_2 and CO_2 exchange. Am J Physiol 224: 909–917
Hinds DS, Calder WA (1971) Tracheal dead space in the respiration of birds. Evolution 25: 429–440
Hinton HE (1966) Plastron respiration. Nature, Lond 209: 220–221
Hochachka PW, Rupert JL, Monge C (1999) Adaptation and conservation of physiological systems in evolution of human hypoxia tolerance. Comp Biochem Physiol A 124: 1–17
Huang W, Ten RT, McLaurie M, Bledsoe G (1996) Morphometry of the human pulmonary vasculature. J Appl Physiol 81: 2123–2133
Huchzermeyer FW, de Ruyk AMC, van Ark H (1988) Broiler pulmonary hypertension syndrome. III. Commercial broiler strains differ in their susceptibility. Onderst J Vet Res 55: 5–9
Hughes GM (1967) Experiments on the respiration of the trigger fish (*Balistes capriscus*). Experientia 23: 1077
Hughes GM (1970) Ultrastructure of the air-breathing organs of some lower vertebrates. 7th Int Congr Electron Microsc, Grenoble 7: 599–600
Hughes GM (1976) Fish respiratory physiology. In: Spencer-Davies P (ed) Perspectives in environmental biology. Pergamon, Oxford, pp 235–245
Hughes GM, Morgan M (1973) The structure of fish gills in relation to their respiratory function. Biol Rev 48: 419–475
Hughes GM, Weibel ER (1976) Morphometry of fish lungs. In: Hughes GM (ed) Respiration of amphibious vertebrates. Academic Press, London, pp 213–232
Joeng H., Tombor B, Albert R., Oltval ZN, Barabási AL (2000) The large-scale organization of metabolic networks. Nature 407: 651–654

Jones JH (1998) Symmorphosis and the mammalian respiratory system: what is optimal design and does it exist? In: Weibel ER, Taylor CR, Bolis L (eds) Principles of animal design: the optimization and symmorphosis debate. Cambridge University Press, Cambridge, pp 241–248
Jones JH, Effmann EL, Schmidt-Nielsen K (1985) Lung volume changes in during respiration in ducks. Respir Physiol 59: 15–25
Joss B, Lighton JRB, Harrison, JF, Suarez RK, Roberts SP (1997) Effects of ambient oxygen tension on flight performance, metabolism and water loss of the honey bee. Physiol Zool 70: 167–174
Joyce GF (1992) Directed molecular evolution. Sci Am 267: 48–55
Julian RJ, Wilson JB (1986) Right ventricular failure as a cause of ascites in broiler and roaster chickens. Proc IVth Int Symp Vet Lab Diag 1986: 608–611
Jürgens JD, Bartels H, Bartels R (1981) Blood oxygen transport and organ weight of small bats and small nonflying mammals. Respir Physiol 45: 243–260
Karsdorp VHM, Dirks BK, van der Linden JC, van Vugt JMG, Baak JPA, van Geijn HP (1996) Placenta morphology and absent or reversed end diastolic flow velocities in the umbilical artery: a clinical and morphometrical study. Placenta 17: 393–399
Kaufman SL (1980) Cell proliferation in the mammalian lung. Int Rev Exp Pathol 22: 131–191
Kleiber M (1961) The fire of life: an introduction to animal energetics. New York, Wiley
Konarzewski M, Gavin A, McDevitt DE, Wallis, IR (2000) Metabolic and organ mass responses to selection for high growth rates in the domestic chicken (*Gallus domesticus*). Physiol Biochem Zool 73: 237–248
König MF, Lucocq JM, Weibel ER (1993) Demonstration of pulmonary vascular perfusion by electron and light microscopy. J Appl Physiol 75: 1877–1883
Krogh A (1941) The comparative physiology of respiratory mechanisms. University of Pennsylvania Press, Philadelphia
Laitman TJ, Reidenberg JS, Marquez S, Gannon PJ (1996) What the nose knows: new understandings of the Neanderthal upper respiratory tract. Proc Natl Acad Sci USA 93: 10543–10545
Liem KF (1987) Form and function of lungs: The evolution of air breathing mechanisms. Am Zool 28: 739–759
Lindstedt SL, Hokanson JF, Wells DJ, Swain SD, Hoppelar H, Navarro V (1991) Running energetics in the pronghorn antelope. Nature, Lond 353: 748
Longo LD, Power GG, Forster RE (1969) Placental diffusing capacity for carbon monoxide at varying partial pressures of oxygen. J Appl Physiol 26: 360–370
Loudon C (1989) Tracheal hypertrophy in mealworms: design and plasticity in oxygen supply systems. J Exp Biol 147: 217–235
Maarek JMI, Chang HK (1991) Pulsatile pulmonary microvascular pressure measured with vascular occlusion techniques. J Appl Physiol 70: 998–1005
Maarek JMI, Grimbert F (1994) Segmental pulmonary vascular resistances during oleic acid lung injury in rabbits. Respir Physiol 98: 179–191
Macklem PT, Bouverot P, Scheid P (1979) Measurement of the distensibility of the parabronchi in duck lungs. Respir Physiol 38: 23–35
Maina JN (1982) A scanning electron microscopic study of the air and blood capillaries of the lung of the domestic fowl (*Gallus domesticus*). Experientia 35: 614–616
Maina JN (1985) A scanning and transmission electron microscopic study of the bat lung. J Zool, Lond 205B: 19–27
Maina JN (1986) The structural design of the bat lung. Myotis 23: 71–77
Maina JN (1987a) The morphology of the lung of the African lungfish, *Protopterus aethiopicus*: a scanning electron microscopic study. Cell Tissue Res 250: 191–196
Maina JN (1987b) The morphology and morphometry of the adult normal baboon lung, *Papio anubis*. J Anat 150: 229–245
Maina JN (1988) The morphology and morphometry of the normal lung of the adult vervet monkey *Cercopithecus aethiops*. Am J Anat 183: 258–267
Maina JN (1989a) A scanning and transmission electron microscopic study of the tracheal air-sac system in a grasshopper (*Chrotogonus senegalensis*, Kraus) – (Orthoptera: Acrididae: Pygomorphinae). Anat Rec 223: 393–405
Maina JN (1989b) Morphometrics of the avian lung. In: King AS, McLelland J (eds) Form and function in Birds, vol 4. Academic Press, London, pp 307–368

Maina JN (1989c) The morphology of the lung of the East African tree frog *Chiromantis petersi* with observations on the skin and the buccal cavity as secondary gas exchange organs: a TEM and SEM study. J Anat 165: 29–43
Maina JN (1989d) The morphology of the lung of the black mamba *Dendroaspis polylepis* (Reptilia: Ophidia: Elapidae): a scanning and transmission electron microscopic study. J Anat 167: 31–46
Maina JN (1989e) The morphology of the lung of a tropical terrestrial slug, *Trichotoxon copleyi* (Mollusca: Gastropoda; Pulmonata): A scanning and transmission electron microscopic study. J Zool, Lond 217: 335–366
Maina JN (1990a) A study of the morphology of the gills of an extreme alkalinity and hyperosmotic adapted teleost *Oreochromis alcalicus grahami* (Boulenger) with particular emphasis on the ultrastructure of the chloride cells and their modifications with water dilution. Anat Embryol 181: 83–98
Maina JN (1990b) The morphology of the gills of the African fresh water crab *Potamon niloticus* (Ortmann-Crustacea-Brachyura-Potamonidae): A scanning and transmission electron microscopic study. J Zool, Lond 221: 499–515
Maina JN (1991) A morphometric analysis of chloride cells in the gills of the teleosts *Oreochromis alcalicus* and *Oreochromis niloticus* and a description of presumptive urea excreting cells in *Oreochromis alcalicus*. J Anat 175: 131–145
Maina JN (1993) Morphometries of the avian lung: the structural-functional correlations in the design of the lungs of birds. Comp Biochem Physiol 105: 397–410
Maina JN (1994) Comparative pulmonary morphology and morphometry: the functional design of respiratory systems. In: Gilles R (ed) Advances in comparative and environmental physiology, vol 20. Springer-Verlag, Heidelberg, pp 111–232
Maina JN (1996) Perspectives on the structure and function in birds. In: Rosskoff E (ed) Diseases of cage and aviary birds. Williams and Wilkins, Baltimore, pp 163–256
Maina JN (1998a) The gas exchangers: structure, function, and evolution of the respiratory processes. Springer-Verlag, Berlin
Maina JN (1998b) The lungs of the volant vertebrates, birds and bats: How are they relatively structurally optimized for this elite mode of locomotion? In: Weibel ER, Taylor CR, Bolis L (eds) Principles of animal design: the optimization and symmorphosis debate. Cambridge University Press, New York, pp 177–195
Maina JN (2000a) Is the sheet-flow design a 'frozen core' (a Bauplan) of the gas exchangers? Comparative functional morphology of the respiratory microvascular systems: illustration of the geometry and rationalization of the fractal properties. Comp Biochem Physiol A 126: 491–515
Maina JN (2000b) Comparative respiratory morphology: themes and principles in the design and construction of the gas exchangers. Anat Rec 261: 25–44
Maina JN (2000c) What it takes to fly: the novel respiratory structural and functional adaptations in birds and bats. J Exp Biol 203: 3045–3064
Maina JN (2000d) Functional morphology of the gas-gland cells of the air-bladder of *Oreochromis alcalicus grahami* (Teleostei: Cichlidae): an ultrastructural study of a fish adapted to a severe, highly alkaline environment. Tissue Cell 32: 117–132
Maina JN, Africa M (2000) Inspiratory aerodynamic valving in the avian lung: functional morphology of the extrapulmonary primary bronchus. J Exp Biol 203: 2865–2876
Maina JN, Cowley H (1998) Ultrastructural characterization of the pulmonary cellular defenses in the lung of a bird, the rock dove, *Columba livia*. Proc Roy Soc, Lond B 265: 1567–1572
Maina JN, Gils van P (2001) Characterization of the airway and vascular systems of the lung of the domestic pig, *Sus scrofa*: Qualitative and quantitative comparison of the topographical, topological and fractal designs of the bronchial, arterial and venous systems. Comp. Biochem Physiol B (in press)
Maina JN, King AS (1982) The thickness of the avian blood-gas barrier: qualitative and quantitative observations. J Anat 134: 553–562
Maina JN, King AS (1984) The structural functional correlation in the design of the bat lung. A morphometric study. J Exp Biol 111: 43–63
Maina JN, King AS, King DZ (1982) A morphometric analysis of the lungs of a species of bat. Respir Physiol 50: 1–11

Maina JN, King AS, Settle G (1989a) An allometric study of the pulmonary morphometric parameters in birds, with mammalian comparison. Phil Trans R Soc, Lond 326B: 1–57
Maina JN, Kisia SM, Wood, CM, Bergman HL, Narahara AB, Laurent P. Walsh PJ (1996b) The morphology and morphometry of the gills of *O. a. grahami*: an ecomorphological study. Int. J. Salt Lakes Res. 5: 131–156
Maina JN, Maloiy GMO (1985) The morphometry of the lung of the lungfish (*Protopterus aethiopicus*): its structural-functional correlations. Proc R Soc, Lond 244B: 399–420
Maina JN, Maloiy GMO (1988) A scanning and transmission electron microscopic study of the lung of a caecilian *Boulengerula taitanus*. J Zool, Lond 215: 739–751
Maina JN, Maloiy GMO (1986) The morphology of the respiratory organs of the African air-breathing catfish (*Clarias mossambicus*): A light, and electron microscopic study, with morphometric observations. J Zool, Lond 209: 421–445
Maina JN, Maloiy GMO (1998) Adaptations of a tropical swamp worm, *Alma emini*, for subsistence in a H_2S-rich habitat: evolution of endosymbiotic bacteria, sulfide metabolizing bodies, and novel processes of elimination of neutralized sulfide complexes. J Struct Biol 122: 257–266
Maina JN, Maloiy GMO, Wood CM (1998) Respiratory stratagems, mechanisms, and morphology of the 'lung' of a tropical swamp worm, *Alma emini* Mich. (Oligochaeta: Glossoscolecidae): a transmission and scanning electron microscope study, with field and laboratory observations. J Zool, Lond 245: 483–495
Maina JN, Maloiy GMO, Warui CN, Njogu EK, Kokwaro ED (1989b) A scanning electron microscope study of the reptilian lungs: the savanna monitor lizard (*Varanus exanthematicus*) and the pancake tortoise (*Malacochersus tornieri*). Anat Rec 224: 514–522
Maina JN, Nathaniel C (2001) A qualitative and quantitative study of a lung of an ostrich *Struthio camelus*. J Exp Biol (in press)
Maina JN, Thomas SP, Dallas DM (1991) A morphometric study of bats of different size: correlations between structure and function of the chiropteran lung. Phil Trans R Soc, Lond 333B: 31–50
Maina JN, Veltcamp CJ, Henry J (1999) Study of the spatial organization of the gas exchange components of a snake lung – sand boa *Eryx colubrinus* (Reptilia: Ophidia: Colubridae) – by latex casting. J Zool (Lond) 247: 81–90
Maina JN, Wood CM, Narahara A, Bergman HL, Laurent P, Walsh P (1996a) Morphology of the swim (air) bladder of a cichlid teleost: *Oreochromis alcalicus grahami* (Trewavas, 1983), a fish adapted to a hyperosmotic, alkaline and hypoxic environment: a brief outline of the structure and function of the swim bladder. In: Dutta HH, Munshi JSD (eds) Fish morphology: horizon of new research. Science Publishers Inc, Lebanon NH, pp 179–192
Malvin GM (1988) Microvascular regulation of cutaneous gas exchange in amphibians. Am Zool 28: 999–1007
Malvin GM, Hlastala MP (1986) Regulation of cutaneous gas exchange by environmental O_2 and CO_2 in the frog. Respir Physiol 65: 99–111
Mandelbrot BB (1983) The fractal geometry of nature. Freeman, New York
Mangum CP, Lykkeboe G, Johansen K (1975) Oxygen uptake and the role of haemoglobin in the East African swampworm *Alma emini*. Comp Biochem Physiol 52A: 477–482
Marcus H (1937) Lungen. In: Bolk L, Goppert E, Kallius E, Lubosch W (eds), Handbuch der vergleichenden Anatomie der Wirbeltiere, III. Urban and Schwarzenberg, Berlin-Wien, pp 909–1018
Mathieu-Costello O, Szewczak JM, Logemann RB, Agey PJ (1992) Geometry of blood-tissue exchange in bat flight muscle compared with bat hindlimb and rat soleus muscle. Am J Physiol 262: R955–R965
Mayhew TM, Joy CF, Haas JD (1984) Structure-function correlation in the human placenta: the morphometric diffusing capacity for oxygen at term. J Anat 139: 691–708
McCutcheon FH (1964) Organ systems in adaptation: the respiratory system. In: Dill DB, Adolph EF, Wilber CG (eds) Handbook of physiology, sect 4, Adaptation to the environment. American Physiological Society, Washington DC, pp 167–191
McGregor LK, Daniels CB, Nicholas TE (1993) Lung ultrastructure and surfactant-like system of the central Australian netted dragon, *Ctenophoruos nuchalis*. Copeia 1993: 326–333
McMahon TA, Kronauer RE (1976) Tree structures: deducing the principle of mechanical design. J Theor Biol 59: 443–466

Menon JG, Arp AJ (1992) Symbiotic bacteria may prevent sulfide poisoning of the body wall of *Urechis caupo*. Am Zool 32: 59A
Metcalfe J, Bartels H, Moll W (1967) Gas exchange in the pregnant uterus. Physiol Rev 47: 782–838
Miller PL (1974) Respiration - aerial gas transport. In: Rockstein M (ed) The physiology of insects, 2nd ed. Academic Press, New York, pp 346–402
Moalli R, Meyers RS, Jackson DC, Millard RW (1980) Skin circulation of the frog, *Rana catesbeiana*: distribution and dynamics. Respir Physiol 40: 137–148
Nikiforov AI, Schlesinger RB (1985) Morphometric variability of the human upper bronchial tree. Respir Physiol 59: 289–299
Okada O, Presson RG, Kirk KR, Godbey PS, Capen RL, Wagner WW (1992) Capillary perfusion patterns in single alveolar walls. J Appl Physiol 72: 1838–1844
Orgeig S, Barr HA, Nicholas TE (1995) Effect of hyperpnea on the cholesterol to disaturated phospholipid ratio in alveolar surfactant of rats. Exp Lung Res 21: 157–174
Pastor LM (1995) The histology of the reptilian lungs. In: Pastor LM (ed) Histology, ultrastructure and immunohistochemistry of respiratory organs in non-mammalian vertebrates. Secretariado de Publicaciones de la Universidad de Murcia, Spain, pp 131–153
Patra AL (1988) Modeling of the upper and lower respiratory systems. In: Gardner JD, Crapo DE, Massaro EJ (eds) Toxicology of the lung. Raven Press, New York, pp 103–115
Pelster B, Scheid P (1992) Counter-current concentration and gas secretion in the fish swim bladder. Physiol Zool 65: 1–16
Pennycuick CJ (1992) Newtonian rules in biology. Oxford University Press, New York
Perry SF (1989) Mainstreams in the evolution of vertebrate respiratory structures. In: King AS, McLelland J (eds) Form and function in birds, vol 4. Academic Press, London, pp 1–67
Perry SF, Bauer AM, Russell AP, Alston JT, Maloney JE (1989) Lungs of the gecko *Rhacodactylus leachianus* (Reptilia: Gekkonidae): A correlative gross anatomical and light and electron microscopic study. J Morph 199: 23–46
Peters, R.H. (1983) The ecological implications of body size. Cambridge University Press, Cambridge
Petroski H (1985) To engineer is human. St Martin's Press, New York
Petroski H (2000). Human factors. Am Sci 88: 304–308
Phleger CF, Smith DG, Macintyre DH, Saunders BS (1978) Alveolar and saccular lung phospholipids of the anaconda, *Eunectes murinus*. Can J Zool 56: 1009–1013
Pickard WF (1974) Transition regime diffusion and the structure of the insect tracheolar system. J Insect Physiol 20: 947–956
Piiper J, Gatz RN, Crawford EC (1976) Gas transport characteristics in an exclusively skin breathing salamander, *Desmognathus fuscus* (Plethodontidae). In: Hughes GM (ed) Respiration in amphibious vertebrates. Academic Press, London, pp 339–356
Piiper J, Scheid P (1992) Modeling of gas exchange in vertebrate lungs, gills, and skin. In: Wood SC, Weber RE, Hargens AR, Millard RW (eds) Physiological adaptations in vertebrates: respiration, circulation, and metabolism. Marcel Dekker Inc, New York, pp 69–95
Piiper J, Tazawa H, Ar A, Rahn H (1980) Analysis of chorioallantoic gas exchange in the chick embryo. Respir Physiol 39: 273–287
Pinkerton KE, Barry BE, O'Neil JJ, Raus JA, Pratt PC, Crapo JD (1982) Morphologic changes in the lung during the lifespan of Fischer 344 rats. Am J Anat 164: 155-174
Pizarro B, Salas A, Paredes J (1970) Mal de altura en aves. Inst Vet Investg Trop Altura Cuarto Boletin Extraord 1970: 147–151
Pough EH, Heiser JB, McFarland WN (1989) Vertebrate life, 3rd edition. Macmillan Publishing Company, New York
Power JHT, Dryle IR, Davidson K, Nicholas TE (1999) Ultrastructural and protein analysis of surfactant in the Australian lungfish, *Neoceratodus forsteri*: evidence for conservation of composition for 300 million years. J Exp Biol 200: 2543–2500
Prange HD (1996) Respiratory physiology: understanding gas exchange. Chapman and Hall, New York
Randall DJ, Burggren WW, Farrell AP, Haswell MS (1981) The evolution of air-breathing in vertebrates. Cambridge University Press, Cambridge
Randall DJ, Daxboeck C (1984) Oxygen and carbon dioxide transfer across fish gills. In: Hoar WS, Randall DJ (eds) Fish physiology, vol 10, part A. Academic Press, New York, London, pp 263–314

Rantin FT, Kalinin AL, Glass ML, Fernandes MN (1992) Respiratory responses to hypoxia in relation to mode of life of two erythrinid species (*Hoplias malabaricus* and *Hoplias lacerdae*). J Fish Biol 41: 805–812

Richards AG, Anderson TF (1942) Further electron microscope studies on arthropod trachea. J NY Entomol Soc 50: 245–247

Ricklefs RE (1985) Modification of growth and development of muscles in poultry. Poultry Sci 64: 1563–1576

Ross Breeders (1999) Ross 308 broiler performance objectives. Ross Breeders, Newbridge

Roughton FJW, Forster RE (1957) Relative importance of diffusion and chemical reaction rates in determining rate of O_2 exchange of pulmonary membrane and volume of blood in lung capillaries. J Appl Physiol 11: 290–302

Scheid P (1978) Analysis of gas exchange between air capillaries and blood capillaries in the avian lung. Respir Physiol 32: 27–49

Scheid P (1979) Mechanisms of gas exchange in bird lungs. Rev Physiol Biochem Pharmacol 86: 137–186

Scheid P, Piiper J (1972) Cross-current gas exchange in the avian lungs: effects of reversed parabronchial air flow in ducks. Respir Physiol 16: 304–312

Scheuermann DW, Klika, E., Lasseel, DG, Bazantova I, Switka A (1997) An electron microscopic study of the parabronchial epithelium in the mature lung of four bird species. Anat Rec 249: 213–225

Schmitz A, Perry SF (1999) Stereological determination of tracheal volume and diffusing capacity of the tracheal walls in the stick insect *Carausius morosus* (Phasmatodea, Lonchodidae). Physiol Zool 72: 205–218

Schraufnagel DE (1987) Microvascular corrosion casting of the lung. A state-of-the-art review. Scan Microsc 1: 1733–1747

Schraufnagel DE, Mehta D, Harshbarger R, Treviranus K, Wand NS (1986) Capillary remodelling in bleomycin-induced pulmonary fibrosis. Am J Pathol 125: 97–106

Severinghaus JW (1971) Transarterial leakage: a possible mechanism of high altitude pulmonary oedema. In: Porter R, Knight J (eds) High altitude physiology: cardiac and respiratory aspects. Churchill-Livinstone, Edinburgh, pp 61–77

Seymour RS, Balylock AJ (2000) The principle of Laplace and scaling of ventricular wall stress and blood pressure in mammals and birds. Physiol Zool 73: 389–405

Silversides FG, Lefrancois MR, Villneuve P (1997) The effect of strain of broiler on physiological parameters associated with ascites syndrome. Poultry Sci 76: 663–667

Singal S, Handerson R, Horsfield K, Harding K, Cumming G (1973) Morphometry of the human pulmonary arterial tree. Circ Res 33: 190–197

Sinha AK, Gleed RD, Hakim TS, Dobson A, Shannon J (1996) Pulmonary capillary pressure during exercise in horses. J Appl Physiol 80: 1792–1798

Slonim NB, Hamilton LH (1971) Respiratory physiology, 2nd edition. The CV Mosby Company, St Louis

Smith DG, Rapson L (1977) Differences in pulmonary microvascular anatomy between *Bufo marinus* and *Xenopus laevis*. Cell Tissue Res 178: 1–15

Snyder GK (1983) Respiratory adaptations in diving mammals. Respir Physiol 54: 269-294

Sobin SS, Fung YC, Tremer HM, Lindal RG (1979) Distensibility of human pulmonary capillary blood vessels in the interalveolar septa. Microvasc Res 17: S87

Somero GN, Childress JJ, Anderson AE (1989) Transport, metabolism, and detoxification of hydrogen sulfide in animals from sulfide-rich marine environments. CRC Crit Rev Mar Aquatic Sci 1: 591–614

Speckmann EW, Ringer Rk (1963) The cardiac output and carotid and tibial blood pressure of the turkey. Can J Biochem Physiol 41: 2337–2354

Stanislaus M (1937) Untersuchungen an der Kolibrilunge. Z Morphol Oekol, Tiere 33: 261–289

Stark-Vancs V, Bell PB, Hutchison VH (1984) Morphological and pharmacological basis for pulmonary ventilation in *Amphiuma tridactylum*: an ultrastructural study. Cell Tissue Res 238: 1–12

Stebbins GL (1984) [Quoted in] Fremontia. January 1984, p 16

Stillman JH (2000) Evolutionary history and adaptive significance of respiratory structures on the legs of intertidal porcelain crabs, genus *Petrolisthes*. Physiol Zool 73: 86–96

Suarez RK (2000) Energy metabolism during insect flight: biochemical design and physiological performance. Physiol Zool 73: 765–771
Tamura M, Hazeki O, Nioka S, Chance B (1989) In vivo study of tissue oxygen metabolism using optical and nuclear magnetic resonance spectroscopies. Annu Rev Physiol 51: 813–834
Taylor CR, Weibel ER (1981) Design of the mammalian respiratory system. Respir Physiol 11: 1–10
Tenney SM, Remmers JE (1963) Comparative morphology of the lung: diffusing area. Nature, Lond 197: 54–56
Thomas SP (1987) The physiology of bat flight. In: Fenton MB, Racey P, Rayner JMV (eds) Recent advances in the study of bats. Cambridge University Press, Cambridge, pp 75–99
Thompson D'AW (1959) On growth and form, 2nd ed. Cambridge University Press, Cambridge
Timwood KT, Hyde DM, Plopper CG (1987) Lung growth of the turkey, *Meleagris gallopavo*: II. Comparison of two genetic lines. Am J Anat 178: 158–169
Tributsch H (1984) How life learned to live: adaptation in nature. The MIT Press, Cambridge, Mass
Veldhuizen RAW, Nag K, Orgeig S, Possmayer F (1998) The role of lipids in pulmonary surfactant. Biochim Biophys Acta 1408: 90–108
Vergara AG, Hughes GM (1981) Phospholipids in washings from the lung of the frog (*Rana pipiens*). J Comp Physiol 27: 117–120
Vidyadaran MK, King AS, Kassim H (1990) Quantitative comparisons of lung structure of adult domestic fowl and the red-jungle fowl, with reference to broiler ascites. Avian Pathol 19: 51–58
Vogel S (1988) Life's devices: The physical world of animals and plants. Princeton University Press, Princeton, NJ
Wagner G (1998) Complexity matters. Science 279: 1158–1159
Wang N, Banzett RB, Nations CS, Jenkins FA (1992) An aerodynamic valve in the avian primary bronchus. J Exp Zool 262: 441–445
Wangensteen OD, Weibel ER (1982) Morphometric evaluation of chorioallantoic oxygen transport in the chick embryo. Respir Physiol 47: 1–20
Weibel ER (1963) Morphometry of the human lung. Springer-Verlag, Berlin
Weibel ER (1973) Morphological basis of the alveolar-capillary gas exchange. Physiol Rev 53: 419–495
Weibel ER (1984) The pathways for oxygen. Harvard University Press, Harvard, MA
Weibel ER (1986) Functional morphology of lung parenchyma. In: Handbook of physiology. The respiratory system. Mechanics of breathing, sect 3, vol III, chapt 8. American Physiological Society, Bethesda MD, pp 89–111
Weibel ER (1991) Fractal geometry: a design principle for living organisms. Am J Physiol 261: L361–L369
Weibel ER (1996) The structural basis of lung function. In: JB West (ed) Respiratory physiology: people and ideas. Oxford University Press, New York, pp 3–46
Weibel ER, Gomez DM (1962) Architecture of the human lung. Science 137: 577–585
Weibel ER, Taylor CR, Bolis L (eds) (1998) Principles of animal design: the optimization and symmorphosis debate. Cambridge University Press, Cambridge
Weibel ER, Taylor CR, Hoppeler H (1991) The concept of symmorphosis: A testable hypothesis of structure-function relationship. Proc Natl Acad Sci USA 88: 10357–10361
Weibel ER, Taylor CR, O'Neil JJ, Leith DE, Gehr P, Hoppeler H, Langman V, Baudinette RV (1983) Maximal oxygen consumption and pulmonary diffusing capacity: A direct comparison of physiologic and morphometric measurements in canids. Respir Physiol 54: 173–188
Weis-Fogh T (1964a) Functional design of the tracheal system of flying insects as compared with the avian lung. J Exp Biol 41: 207–228
Weis-Fogh T (1964b) Diffusion in insect flight muscle, the most active tissue known. J Exp Biol 41: 229–256
Welsch U (1981) Fine structure and enzyme histochemical observations on the respiratory epithelium of the caecelian lungs and gills. A contribution to the understanding of the evolution of the vertebrate respiratory epithelium. Arch Histol Jap, Okayama 44: 117–133
West BJ (1987) Fractals, intermittency and morphogenesis. In: Degn H, Holden AV, Olsen LF (eds) Chaos in biological systems. Plenum Press, New York, pp 305–314
West BJ (1990) Fractal physiology and chaos in medicine. World Scientific Publications, Singapore
West BJ, Brown JH, Enquist BJ (1999) The fourth dimension of life: Fractal geometry and allometric scaling of organisms. Science 284: 1677–1679

West JB (1974) Respiratory physiology: the essentials. Williams and Wilkins, Baltimore
West JB, Mathieu-Costello O, Jones JH, Birks EK, Logerman RB, Pascoe JR, Tyler WS (1993) Stress failure in pulmonary capillaries in racehorses with exercise induced pulmonary haemorrhage. J Appl Physiol 75: 1097–1109
Wigglesworth VB (1974) Insect physiology, 7th ed. Chapman and Hall, London.
Wilson TA, Bachofen H (1982) A model for mechanical structure of the alveolar duct. J Appl Physiol 52: 1064–1070
Wimsatt WA (1970). Biology of bats. Academic Press, London
Wirtz HRW, Dobbs LG (1990) Calcium mobilization and exocytosis after one mechanical stretch of lung epithelial cells. Science 250: 1266–1269
Woldenberg MJ, Cumming G, Harding K, Horsfield K, Prowse K, Singal S, (1970) Law and order in the human lung. Office of the Naval Research Department (AD 709602). U.S. Department of Commerce, Springfield, VA
Zaccone G, Goniakowska-Witalinska L, Lauweryns JM, Fasulo S, Tagliafierro G (1989) Fine structure and serotonin immunohistochemistry of the neuroendocrine cells in the lungs of the bichirs *Polpterus delhezi* and *P. ornatipinnis*. Bas Appl Istochem 33: 277–294

Figures

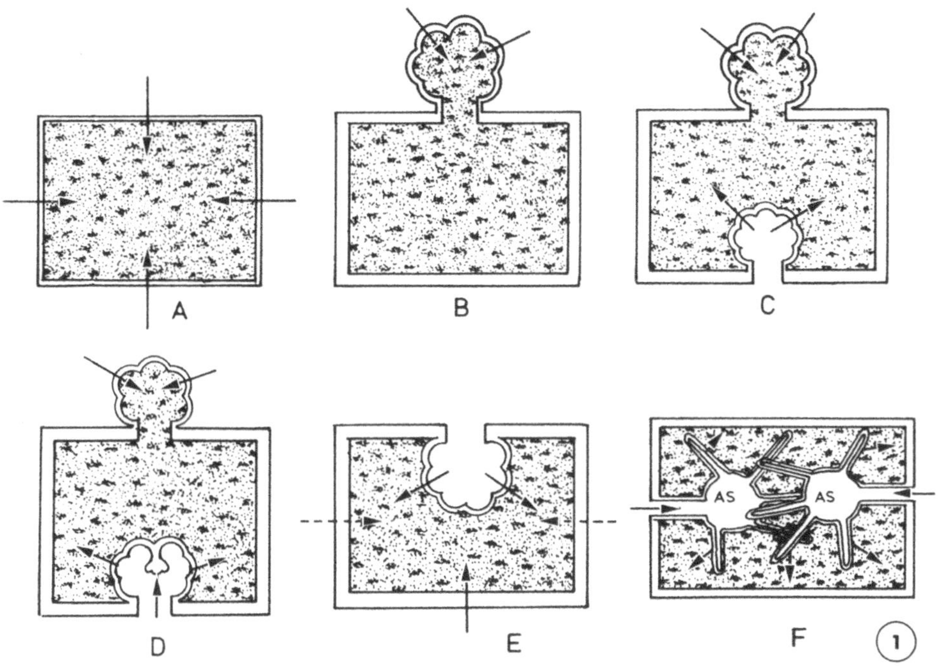

Fig. 1A–F. The basic designs of the gas exchangers. The simplest design (**A**) is found in the unicells and the very simple organisms. Oxygen transfer (shown with *arrows*) occurs by diffusion across the cell membrane. In the more complex animals, gas exchangers have formed either as evaginations (**B**) or invaginations (**C–F**). **B** Example of an aquatic breather; **C** A bimodal breather with an unmodified accessory respiratory organ; **D** A bimodal breather with a modified accessory respiratory organ; **E** A terrestrial air breather – in amphibians the skin (shown with *dashed arrows*) serves as an important respiratory pathway; and **F** A tracheal system in which air is delivered directly to the tissue cells across the air sacs (*AS*)

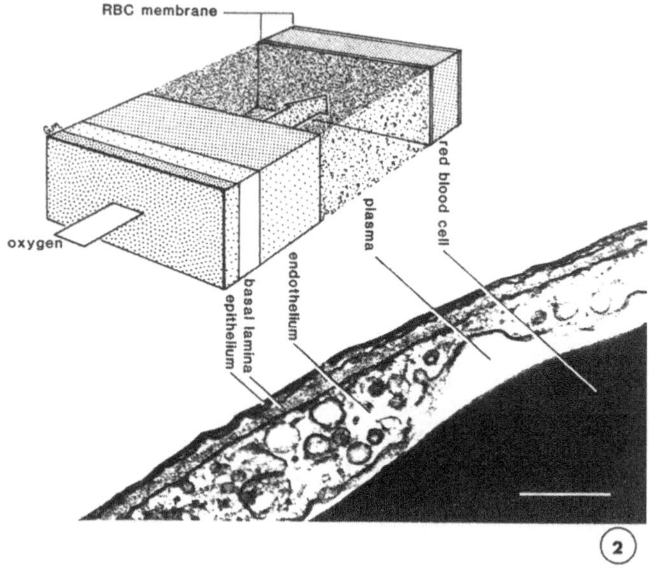

Fig. 2. A schematic diagram showing the components of the air-haemoglobin pathway, namely, the blood–gas (tissue) barrier, the plasma layer and the red blood cell. An electron micrograph of the lung of black-headed gull, *Larus ridibundus* is shown against the drawing. *Scale bar*, 0.4 μm

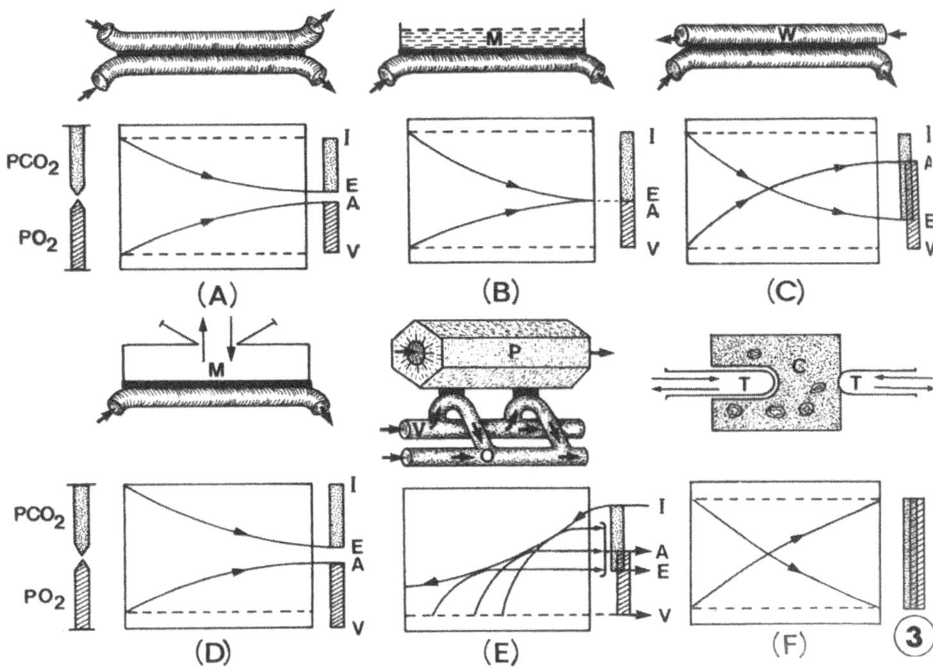

Fig. 3 A–F. Schematic drawings showing the presentations of the respiratory media in different gas exchangers. The Po_2 and Pco_2 profiles in the inspired air (*I*), expired air (*E*), venous blood (*V*) and arterial blood (*A*) show the respiratory efficiencies. **A** A theoretical co-current model in which the respiratory media flow in the same direction. **B** Infinite pool model in which the gas exchanger is perfused but not ventilated, e.g. the skin. **C** Counter-current model in which media flow in opposite directions (e.g. fish gills) – the Po_2 in the arterial blood exceeds that in the end expired air. **D** Ventilated (uniform) pool model in which the gas exchanger is perfused and ventilated (e.g. the mammalian lung). **E** Cross-current model where air and blood flow cross perpendicularly (e.g. the avian lung) – the Po_2 in the arterial blood may exceed that in the expired air. **F** The insectan tracheal model in which O_2 is delivered directly to the tissue cells by the trachea (*T*). Theoretically, the Po_2 in the arterial blood (body tissues in the case of insects) may exceed that in the expired respiratory media only in models **C**, **E** and **F**. *m*, respiratory medium (air/water); *p*, parabronchus; *c*, cell; *v*, blood (venous); *w*, water. In the schematic drawings, the *single arrows* show the directions of flow of respiratory media and the diffusion of O_2; the *parallel arrows* in model **D** show the tidal ventilation of the gas exchanger

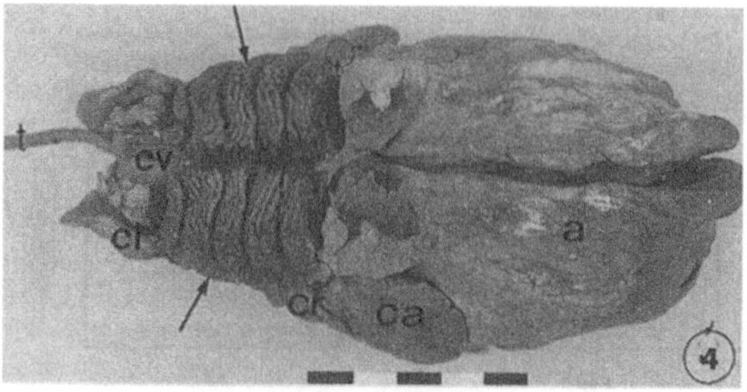

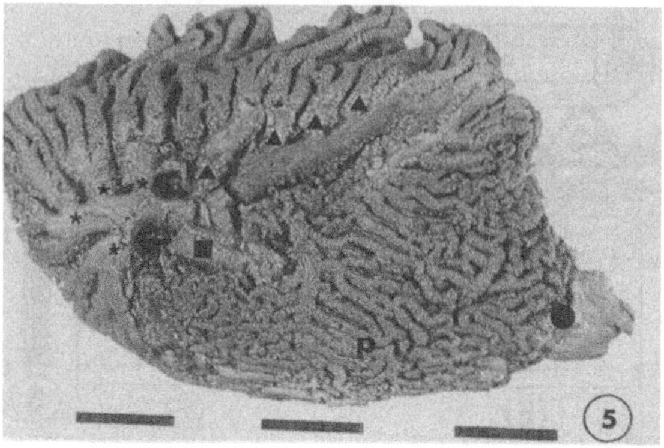

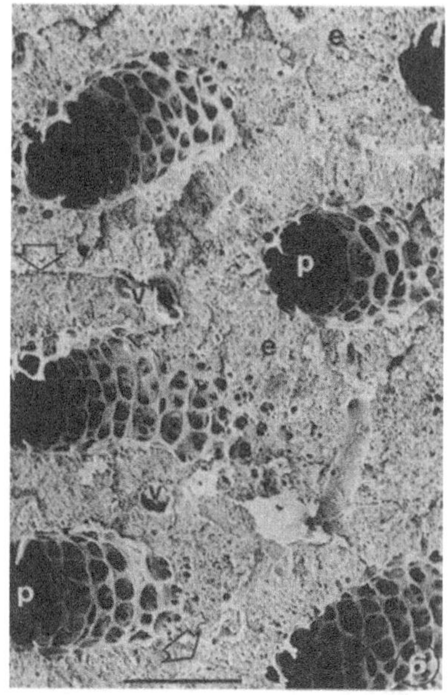

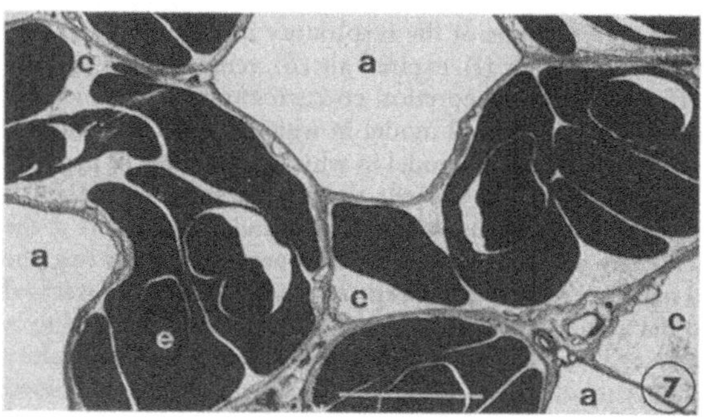

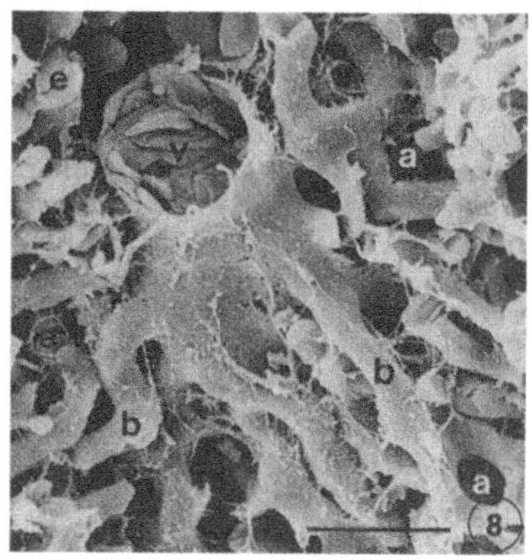

Fig. 4. The lung-air sac system of the domestic fowl, *Gallus gallus* var. *domesticus* showing the lung (*arrows*) intercalated between the cranial air sacs [cervical air sacs (*cv*), clavicular air sacs (*cl*) and craniothoracic air sacs (*cr*)] and the caudal thoracic air sacs [caudal thoracic air sacs (*ca*) and abdominal air sacs (*a*)]. *t*, Trachea; ❍, ostia. *Scale bar*, 10 mm

Fig. 5. Medial view of the lung of the domestic fowl, *Gallus gallus* var. *domesticus*, showing the medioventral secondary bronchi (*stars*), mediodorsal secondary bronchi (▲), primary bronchus (■), parabronchi (*p*) and an ostium (●). *Scale bar*, 10 mm

Fig. 6. Lung of the emu, *Dromiceius novaehollandiae*, showing parabronchial lumina (*p*) surrounded by exchange tissue (*e*), interparabronchial septa (*arrows*) and interparabronchial blood vessels (*v*). *Scale bar*, 0.2 mm

Fig. 7. Air capillaries (*a*) and blood capillaries (*c*) containing red blood cells (*e*) of exchange tissue of the lung of *Gallus gallus* var. *domesticus*. *Scale bar*, 0.4 μm

Fig. 8. An intraparabronchial blood vessel (*v*) of the lung of the emu, *Dromiceius novaehollandiae*, giving rise to blood capillaries (*b*) that interdigitate with air capillaries (*a*). *e*, erythrocytes. *Scale bar*, 4 μm

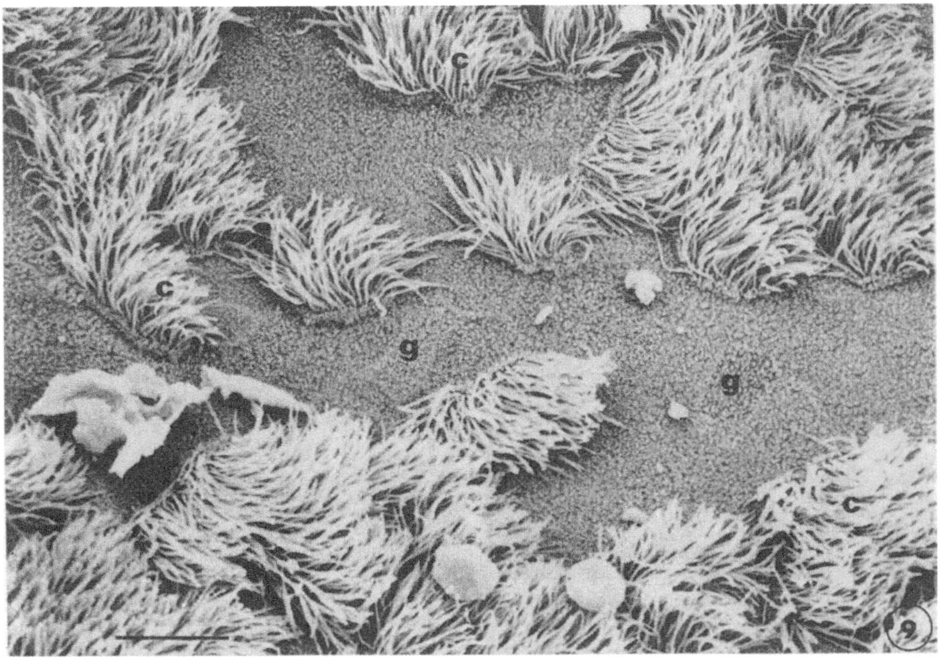

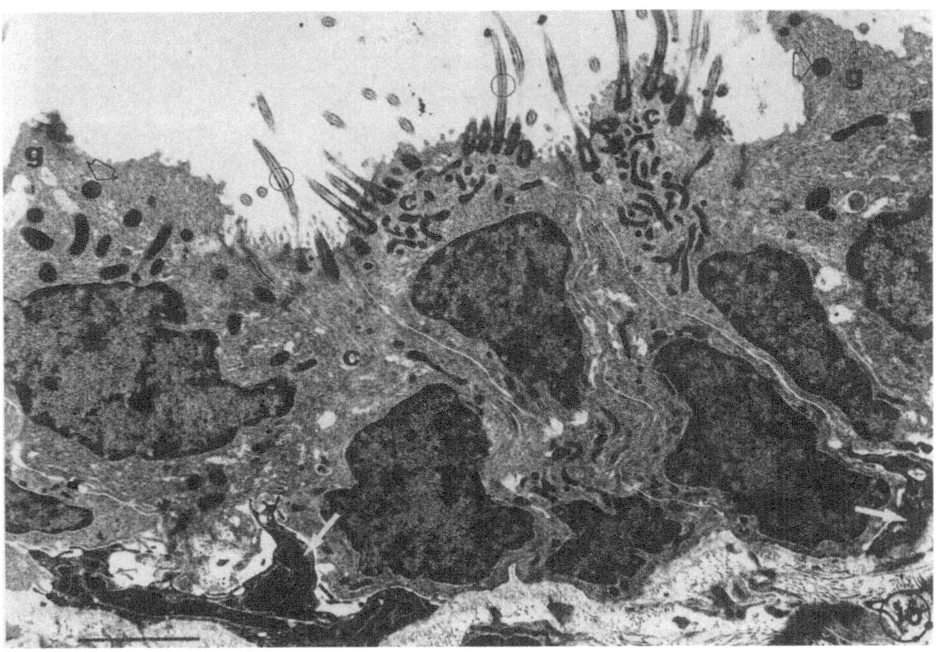

Figs. 9, 10. Epithelial lining of the trachea of the lung of the lesser bushbaby, *Galago senegalensis*, showing ciliated cells (*c*) and nonciliated, mucus secretory cells (*g*). *Open arrows*, secretory granules; *white arrows*, basal cells; *circles*, cilia. **9** *Scale bar*, 9 μm; **10**, 3 μm

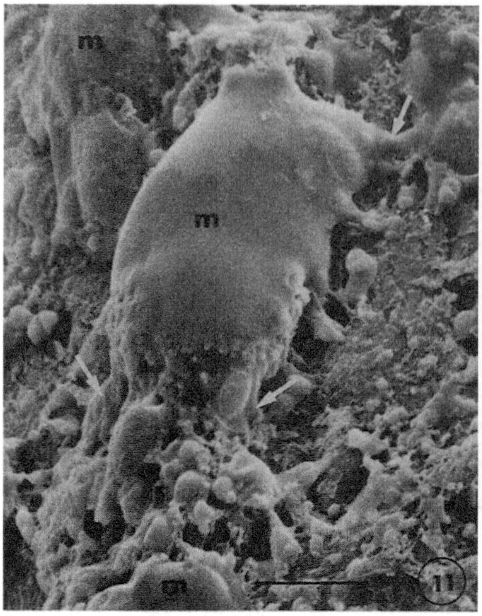

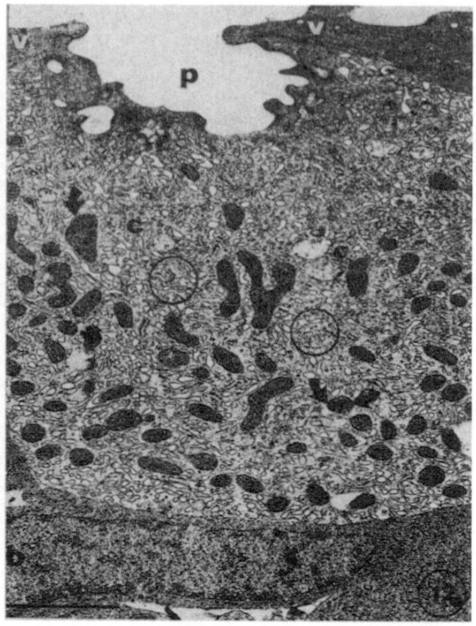

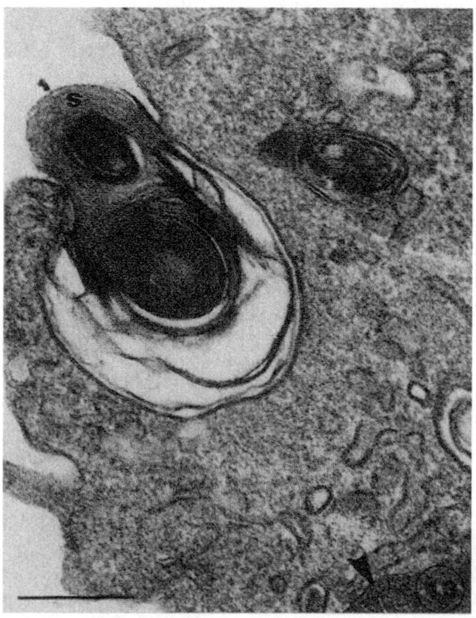

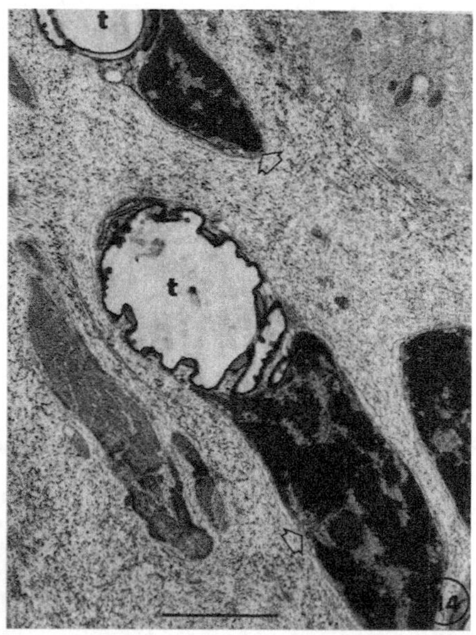

Fig. 11. Macrophages (*m*) on the surface of the lung of the tree frog, *Chiromantis petersi. Arrows,* filopodia. *Scale bar,* 6 μm

Fig. 12. Chloride cell of the gills of *Oreochromis alcalicus grahami* showing numerous polymorphic mitochondria (*arrows*) interspersed by dense microtubular network (O). *p*, pore; *v*, pavement cells; *b*, basal cell. *Scale bar,* 2 μm

Fig. 13. Type I cell of the lung of the vervet monkey, *Cercopithecus aethiops,* discharging surfactant (*s*) onto the alveolar surface. *Arrowhead,* mitochondrion. *Scale bar,* 0.4 μm

Fig. 14. Trachea (*t*) in the tissue of the gastrointestinal system of the desert locust, *Locusta migratoria migratoria,* surrounded by tracheoblasts (*arrows*). *Scale bar,* 0.5 μm

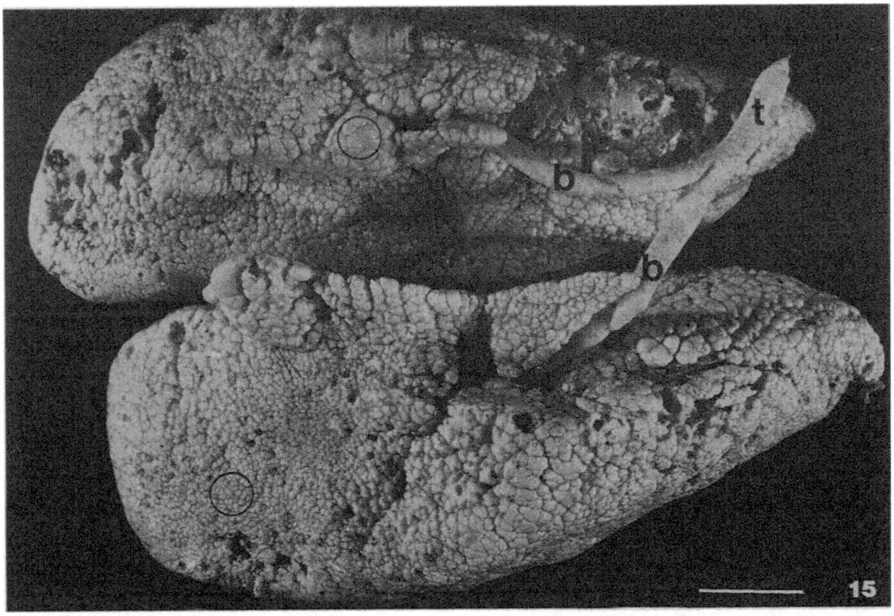

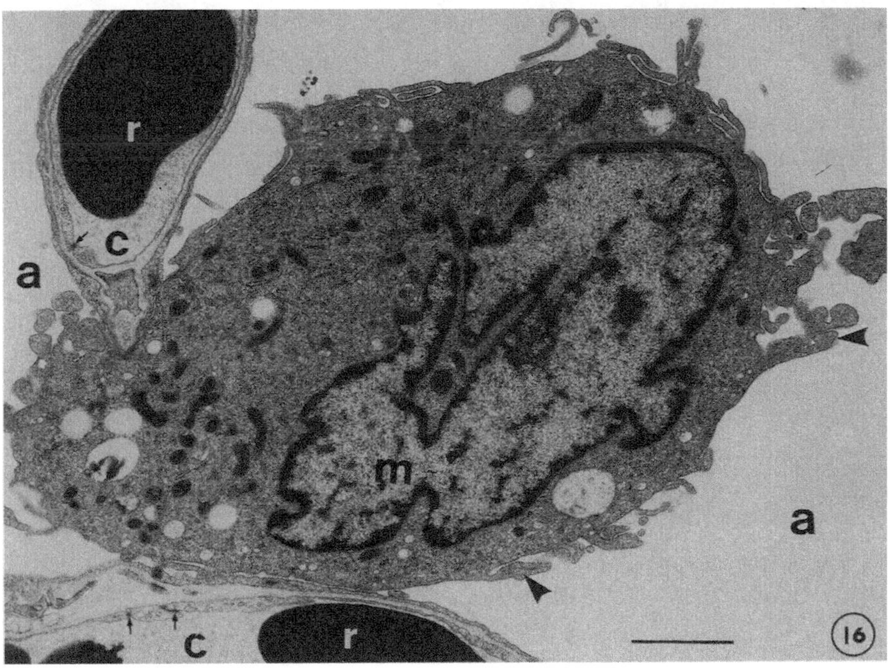

Fig. 15. Medial view of the lung of the monitor lizard, *Varanus exanthematicus*, showing the trachea (*t*) and bronchi (*b*). The terminal gas exchange components greatly vary in size in different parts of the lung (*circles*). *Scale bar*, 2 mm

Fig. 16. An alveolar macrophage (*m*) of the lung of a bat, *Miniopterus minor*, passing through an interalveolar pore. *a*, alveolus; *r*, red blood cell; *c*, blood capillaries; *arrows*, endothelial cell tight junctions; *arrowheads*, filopodia. *Scale bar*, 5 µm

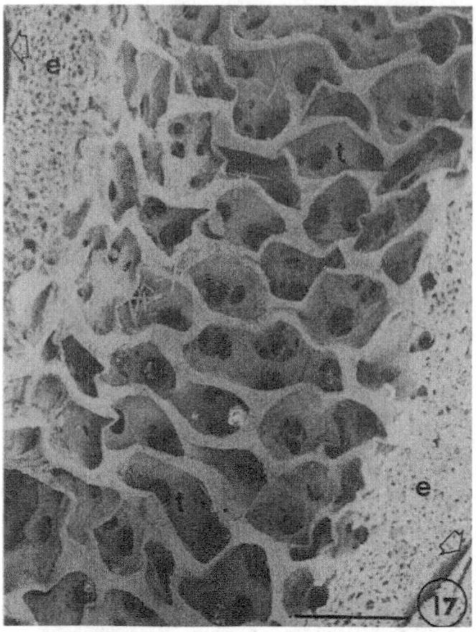

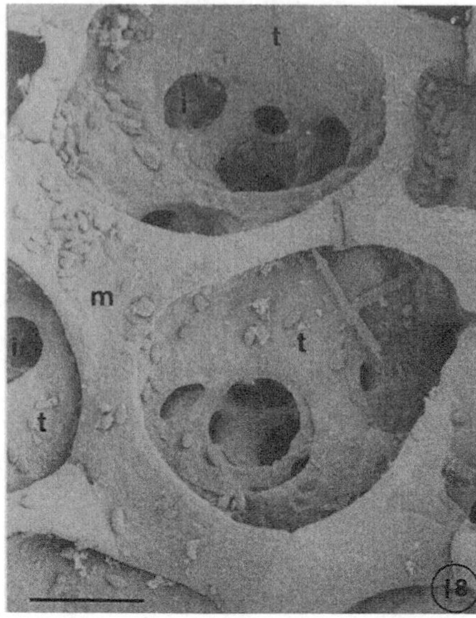

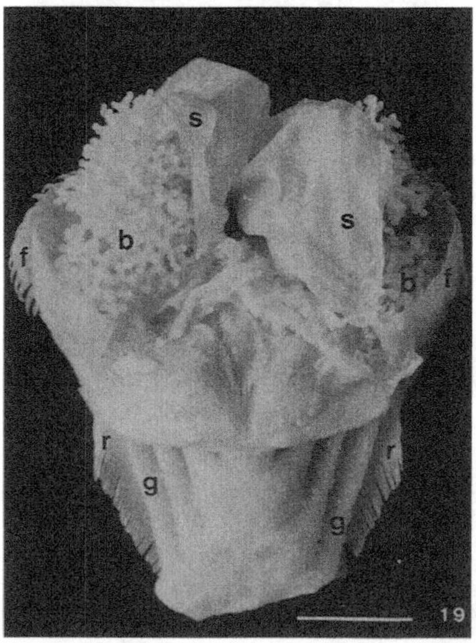

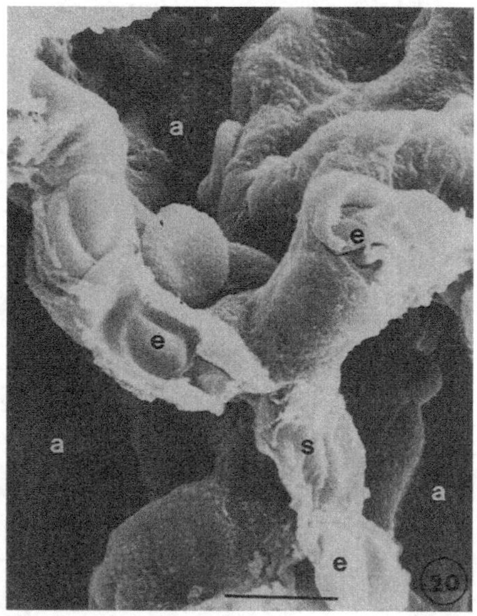

Fig. 17. Longitudinal view of a parabronchus of the lung of the domestic fowl, *Gallus gallus* var. *domesticus,* showing atria (*t*) and exchange tissue (*e*). *Arrows,* interparabronchial septa. *Scale bar,* 0.05 mm

Fig. 18. Close-up of the internal aspect of a parabronchus of the lung of the domestic fowl, *Gallus gallus* var. *domesticus,* showing atria (*t*) and infundibulae (*i*). *m,* Interatrial muscles. *Scale bar,* 0.014 mm

Fig. 19. Respiratory organs of the African catfish, *Clarias mossambicus,* showing gill arches (*g*), labyrinthine organs (*b*) and suprabranchial chamber membrane (*s*). *r,* gill rakers; *f,* gill fans. *Scale bar,* 10 mm

Fig. 20. Interalveolar septum (*s*) of the lung of the vervet monkey, *Cercopithecus aethiops. a,* alveoli; *e,* erythrocytes contained in the blood capillaries; *s,* interalveolar septum. *Scale bar,* 1.5 μm

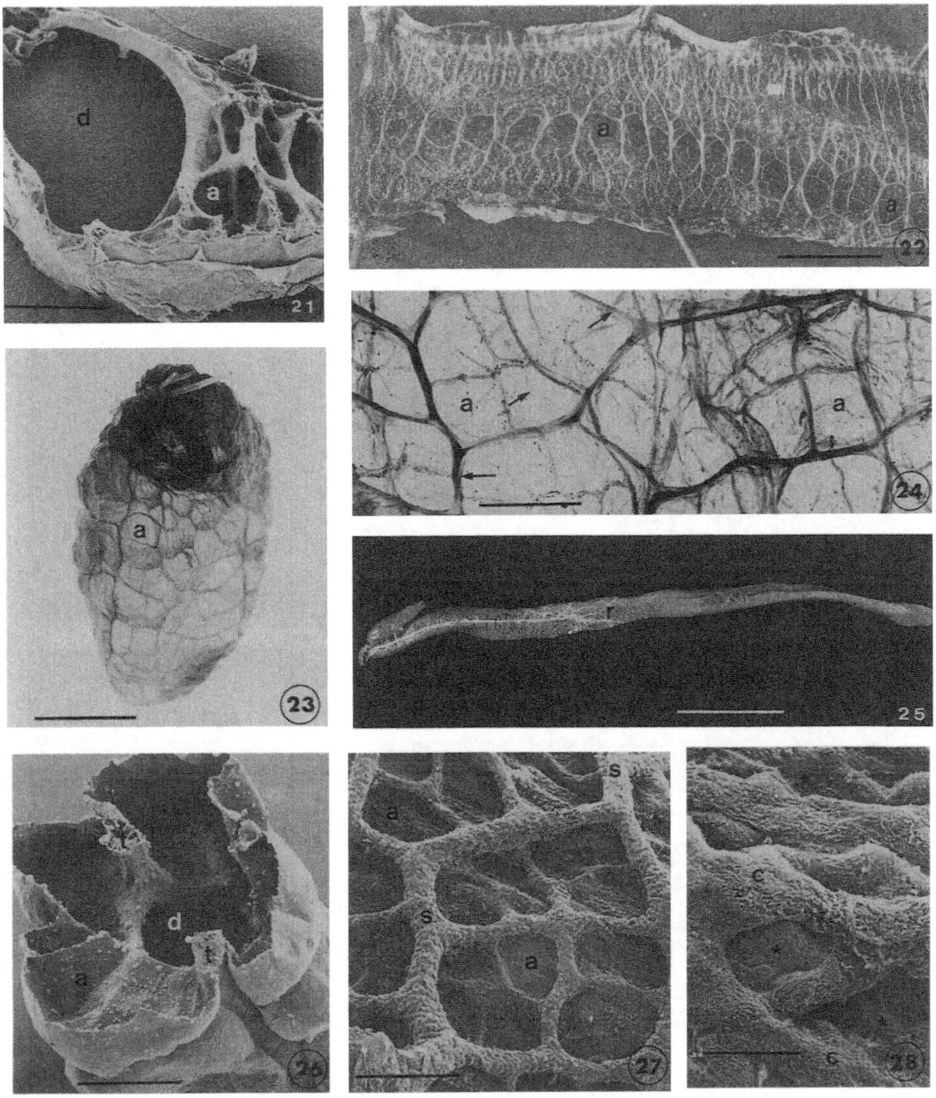

Fig. 21. Transverse view of the lung of the African lungfish, *Protopterus aethiopicus*, showing an eccentrically located air duct (*d*) and gas exchange components (*a*). *Scale bar*, 0.7 mm

Fig. 22. Longitudinal (internal) view of the lung of the African lungfish, *Protopterus aethiopicus*, showing an eccentrically located air duct (*d*) and gas exchange components (*a*). *Scale bar*, 1 mm

Figs. 23, 24. Lung of the tree frog, *Chiromantis petersi*, showing shallow terminal gas exchange components (*a*) formed by septa (*arrows*). *Scale bar*: **23**, 10 mm; **24**, 1.8 mm

Figs. 25, 26. The tubular lung of the caecilian, *Bourengerula taitanus*. Only the right lung (*r*) is well developed. The lung is supported by two diametric trabeculae (*t*). *d*, air duct; *circle*, pulmonary artery; *a*, air cells. *Scale bar*: **25**, 1 mm; **26**, 0.5 mm

Fig. 27. View of the air cells of the lung of the lungfish, *Protopterus aethiopicus*, showing septa (*s*) and terminal gas exchange components (*a*). *Scale bar*, 0.2 mm

Fig. 28. Internal surface of the lung of the slug, *Trichtoxon copleyi*, showing haemolymphatic vessels (*c*) bulging over the respiratory surface. *Stars*, intercapillary depressions. *Scale bar*, 0.2 mm

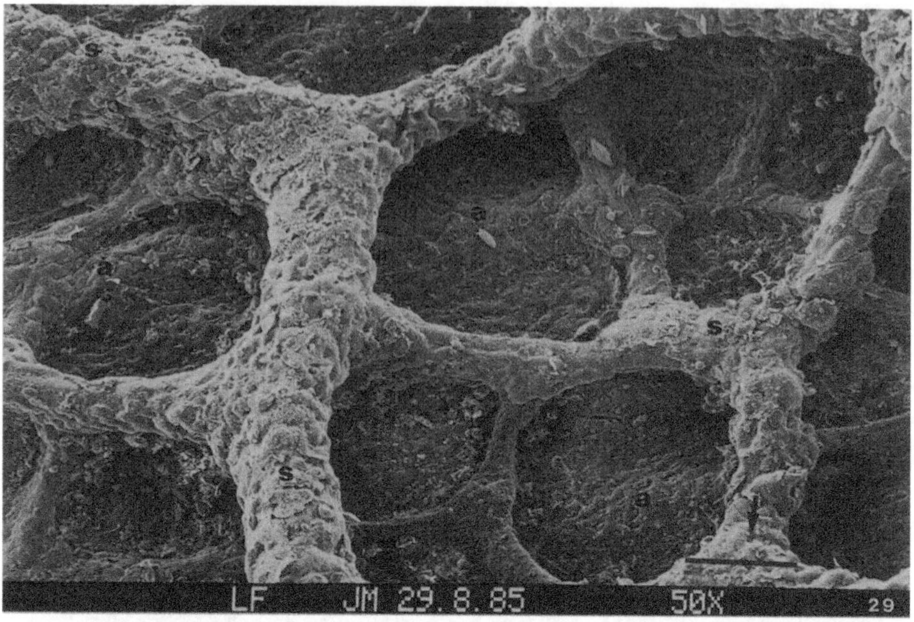

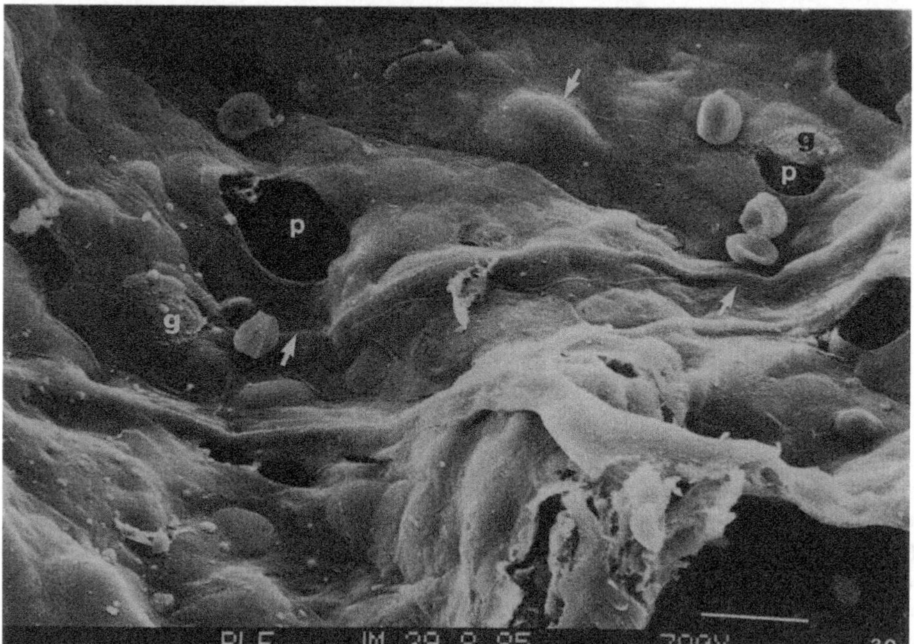

Fig. 29. Septa (*s*) that internally subdivide the lung of the lungfish, *Protopterus aethiopicus*, into two levels of peripheral air cells (*a*). The epithelial cell nuclei (perikarya) are located in the depressions (*arrow*). *Scale bar*, 0.2 mm

Fig. 30. Alveolar surface of the lung of the baboon, *Papio anubis*, showing interalveolar pores (*p*) and granular pneumocytes (*g*). The *arrows* mark the extents (the intercellular junctions of the highly squamous type I pneumocytes). *Scale bar*, 10 µm

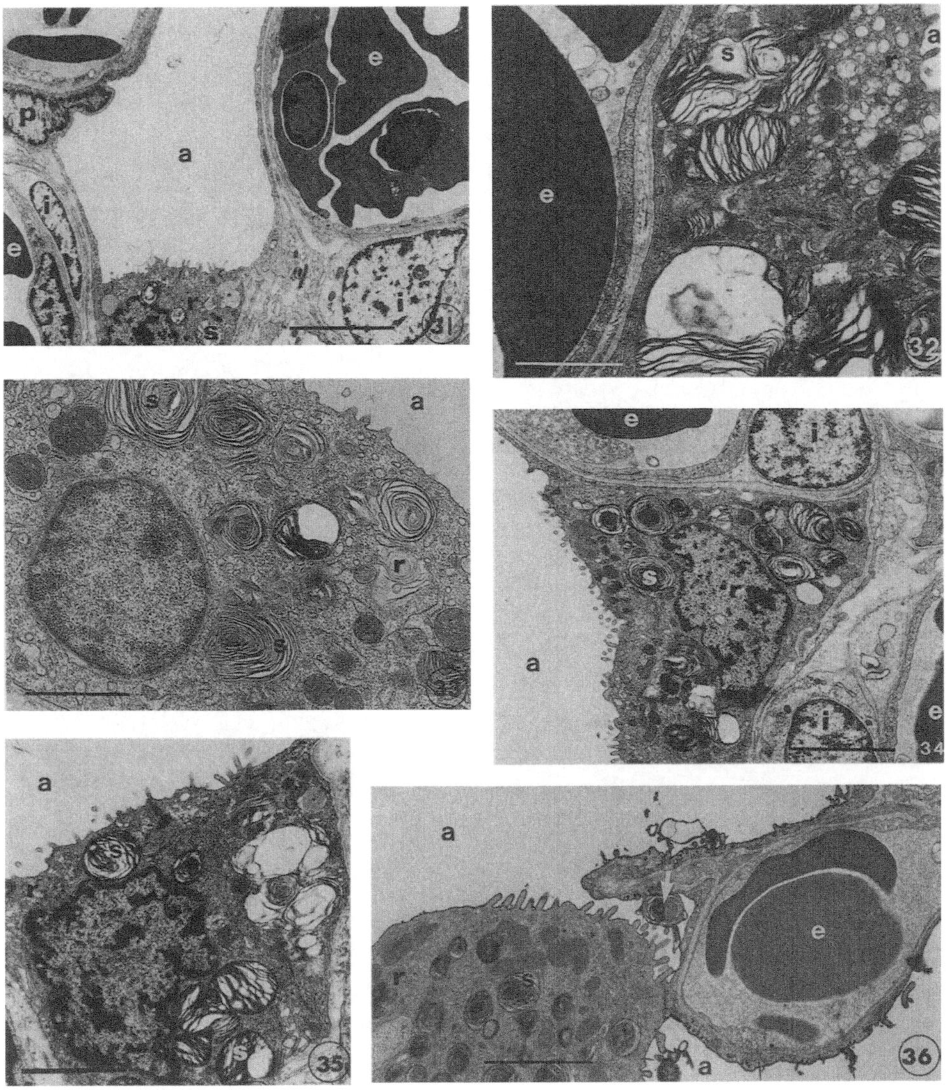

Figs. 31–36. Granular pneumocytes (type II cells) (*r*) of lungs of different vertebrate animals. **31, 34** Tree frog, *Chiromantis petersi*. **32** Lungfish, *Protopterus aethiopicus*. **33** Bat lung, *Miniopterus minor*. **35** Snake, *Dendroaspis polylepis*. **36** Vervet monkey, *Cercopithecus aethiops*. *a*, air space; *e*, red blood cells contained in blood capillaries; *s*, osmiophilic lamellated bodies; *i*, interstitial cells; *p*, epithelial cell; the *arrow* in **36** shows a lamellated body that has been discharged onto the alveolar surface. *Scale bar*: **31**, 4 μm; **32**, 0.5 μm; **33**, 3 μm; **34**, 1 μm; **35**, 0.7 μm; **36**, 2 μm

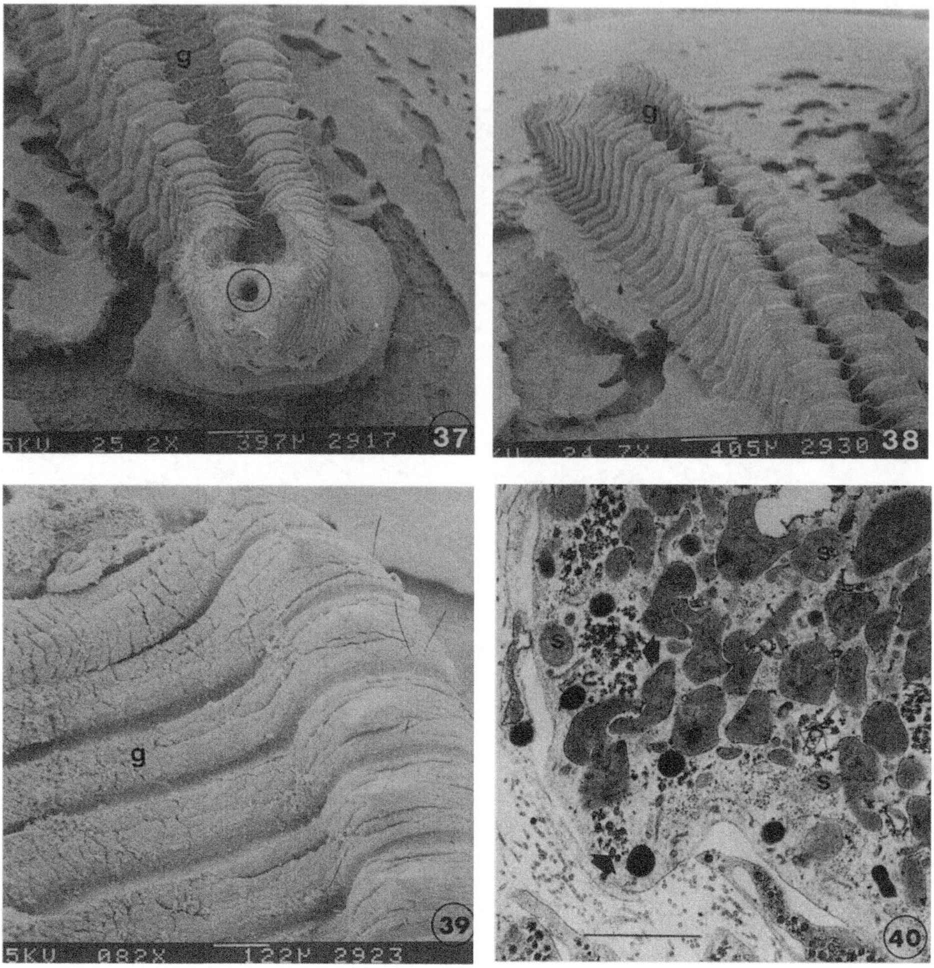

Figs. 37–40. 37–39 The respiratory groove (lung) (*g*) of the swampworm, *Alma emini*. **40** Prokaryotic bacterial symbionts (*s*) contained in the cells that line the respiratory groove of *Alma emini*. *arrows*, putative products of detoxification of sulphur. **37–39** *Scale bars* shown on the figs; **40** *scale bar*, 0.7 μm

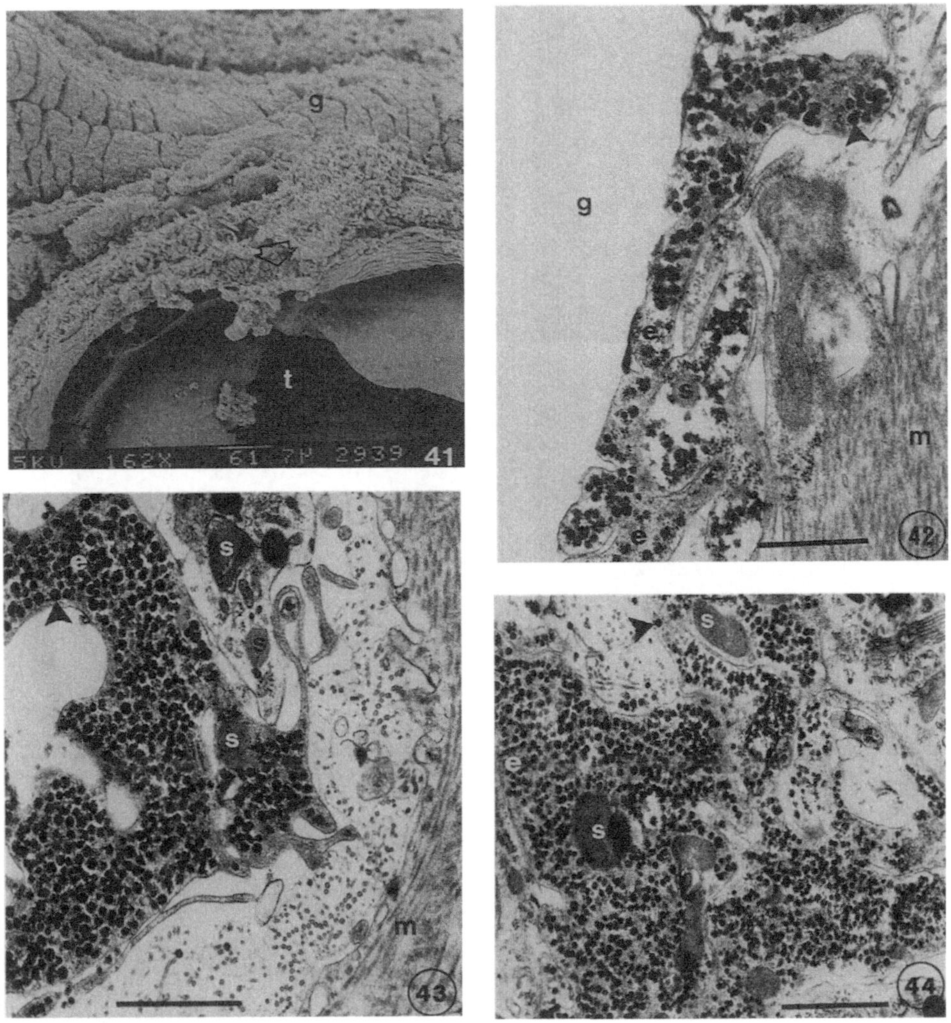

Figs. 41–44. The respiratory groove (*g*) of the swampworm, *Alma emini*. **41** The *arrow* shows blood vessels underlying the epithelium lining the respiratory groove. *t*, gastrointestinal system. **42–44** Epithelial cells (*e*) that line the respiratory groove. *s*, prokaryotic bacterial symbionts; *arrowheads*, metabolites of sulphur detoxification; m, muscle tissue. *Scale bar*: **41**, 61.7 µm; **42**, 0.6 µm; **43**, 0.7 µm; **44**, 0.9 µm

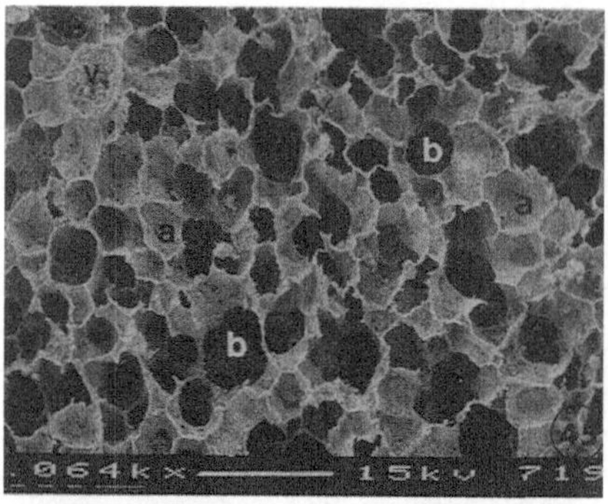

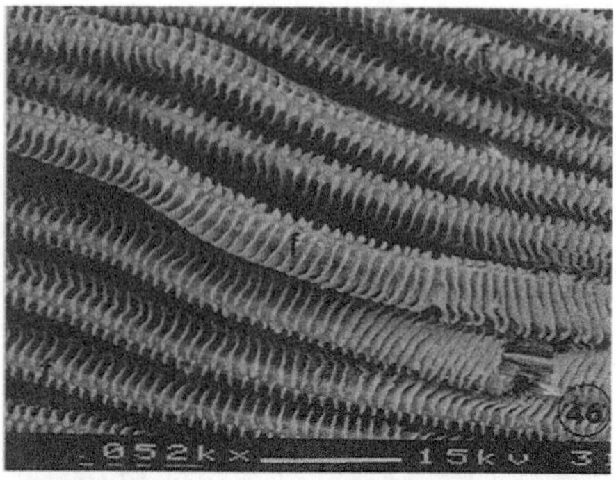

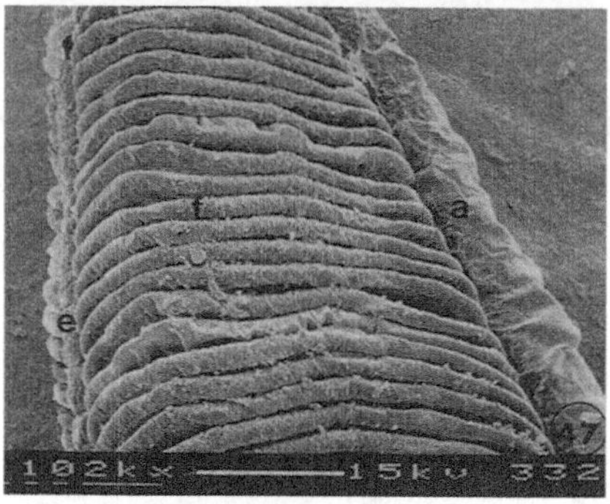

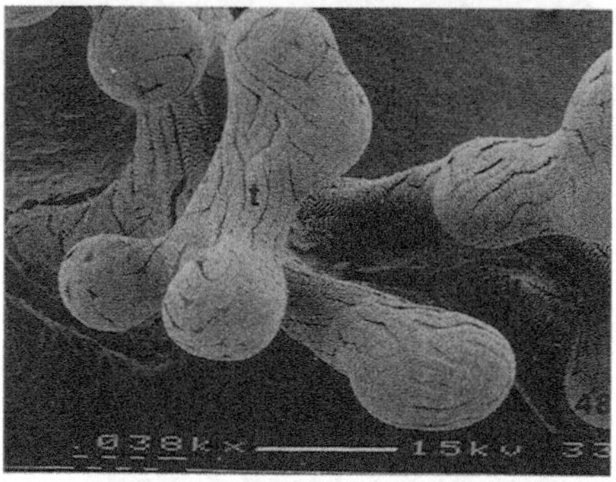

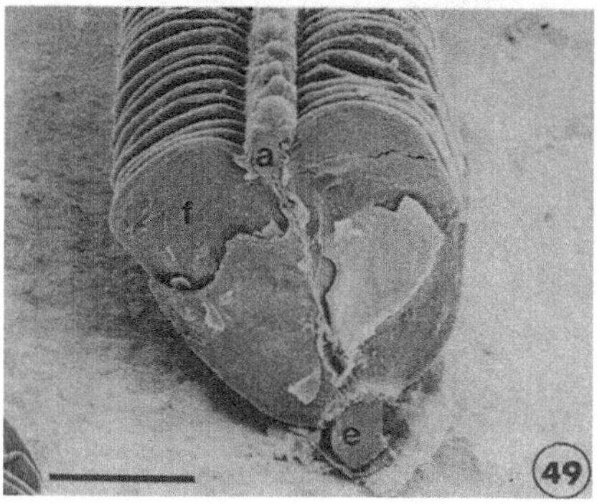

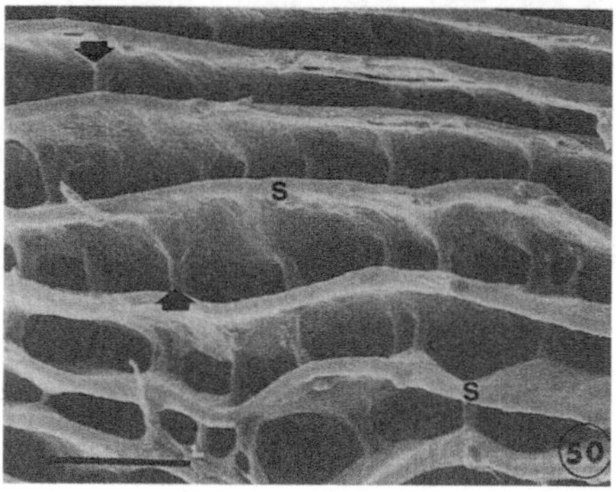

Fig. 45. Lung of the vervet monkey, *Cercopithecus aethiops,* showing alveoli (*a*) and respiratory bronchioles (*b*). *v*, blood vessel. *Scale bar*, 0.3 mm

Fig. 46. Gill filaments (*k*) of a branchial arch of *Oreochromis alcalicus grahami.* Secondary lamellae are seen as bilateral hemispherical flaps that extend to opposite sides of the gill filaments. *Scale bar*, 0.4 mm

Fig. 47. Longitudinal view of a gill arch of the fresh water crab, *Potamon niloticus,* showing filaments (*f*). *a*, afferent vascular channel; *e*, efferent vascular channel. *Scale bar*, 0.2 mm

Fig. 48. Labyrinthine organ of the catfish, *Clarias mossambicus,* showing vascular tracts (*t*) that cover the organ. *Scale bar*, 0.5 mm

Fig. 49. A transverse view of the gill arch of the fresh water crab, *Potamon niloticus,* showing gill filaments (*f*) and afferent (*a*) and efferent vessels (*e*). *Scale bar*, 0.3 mm

Fig. 50. Book lung of a desert scorpion, *Paruroctonus mesaensis,* showing lamellae (*s*) separated by vertical struts (*arrows*) that prevent their collapse. *Scale bar*, 0.4 mm. (Reproduced courtesy of Prof. R.D. Farley)

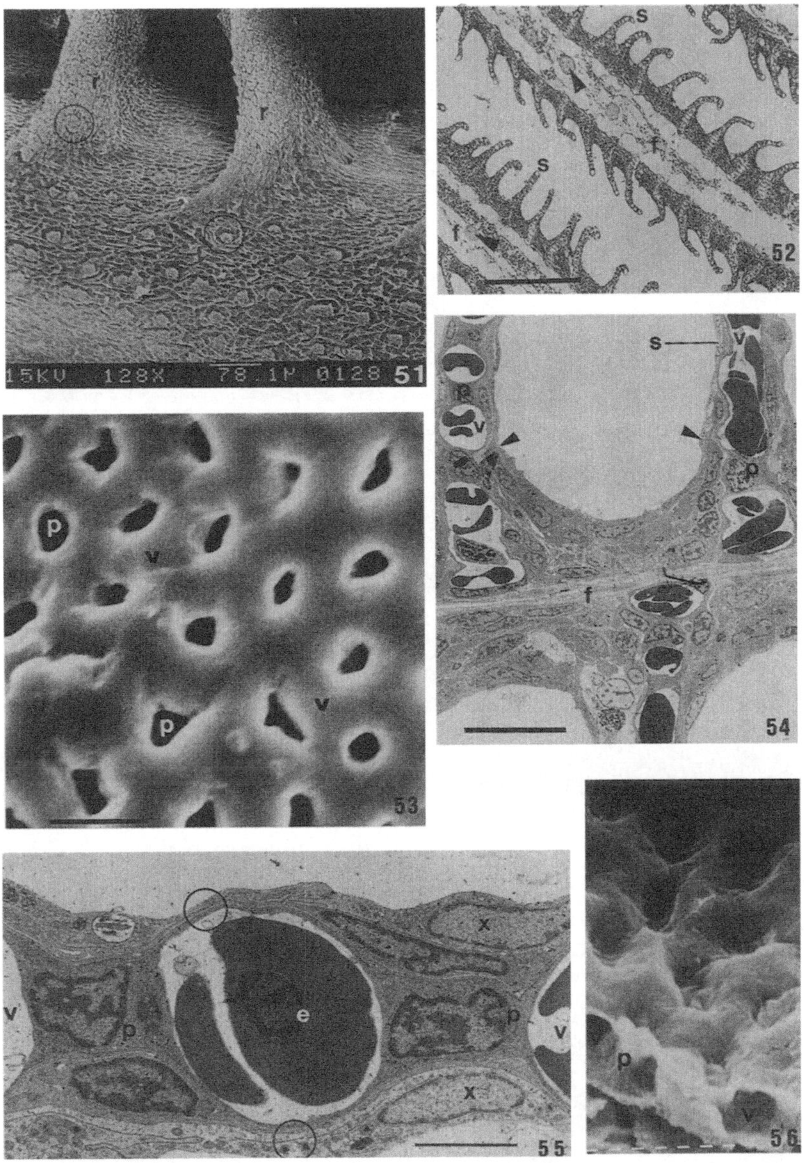

Fig. 51–56. Fish gills of *Oreochromis alcalicus grahami*. **51** A gill arch with gill rakers (*r*) extending from the dorsal surface. Goblet and mitochondria cells (*circles*) can be seen on the epithelial surface. *Scale bar*, 78 µm. **52** Gill filaments (*f*) and secondary lamellae (*s*). The vascular channels in the secondary lamellae can be seen in the secondary lamellae. *Arrowheads*, cartilaginous plates that provide support to the gill filaments. *Scale bar*, 50 µm. **53** Cast of a secondary lamella showing vascular channels (*v*) separated by pillar cells (*p*). *Scale bar*, 3 µm. **54** Cross section of a gill filament (*f*) showing secondary lamellae (*s*) originating from opposite sides. *p*, pillar cells; *v*, vascular channels. The *arrowheads* mark the junction between the elaborate primary epithelium that is found on the gill filament and the interlamellar space and the thin secondary epithelium that lines the secondary epithelium. *Scale bar*, 43 µm. **55** Secondary epithelium lining a secondary lamella showing vascular channels (*v*) containing red blood cells (*e*). *p*, pillar cells; *x*, epithelial cell; *circle*, water-blood barrier. *Scale bar*, 14 µm. **56** A view of the pillar cells (*p*) (with epithelial cell detached). *v*, vascular channels. *Scale bars*, 216 mm

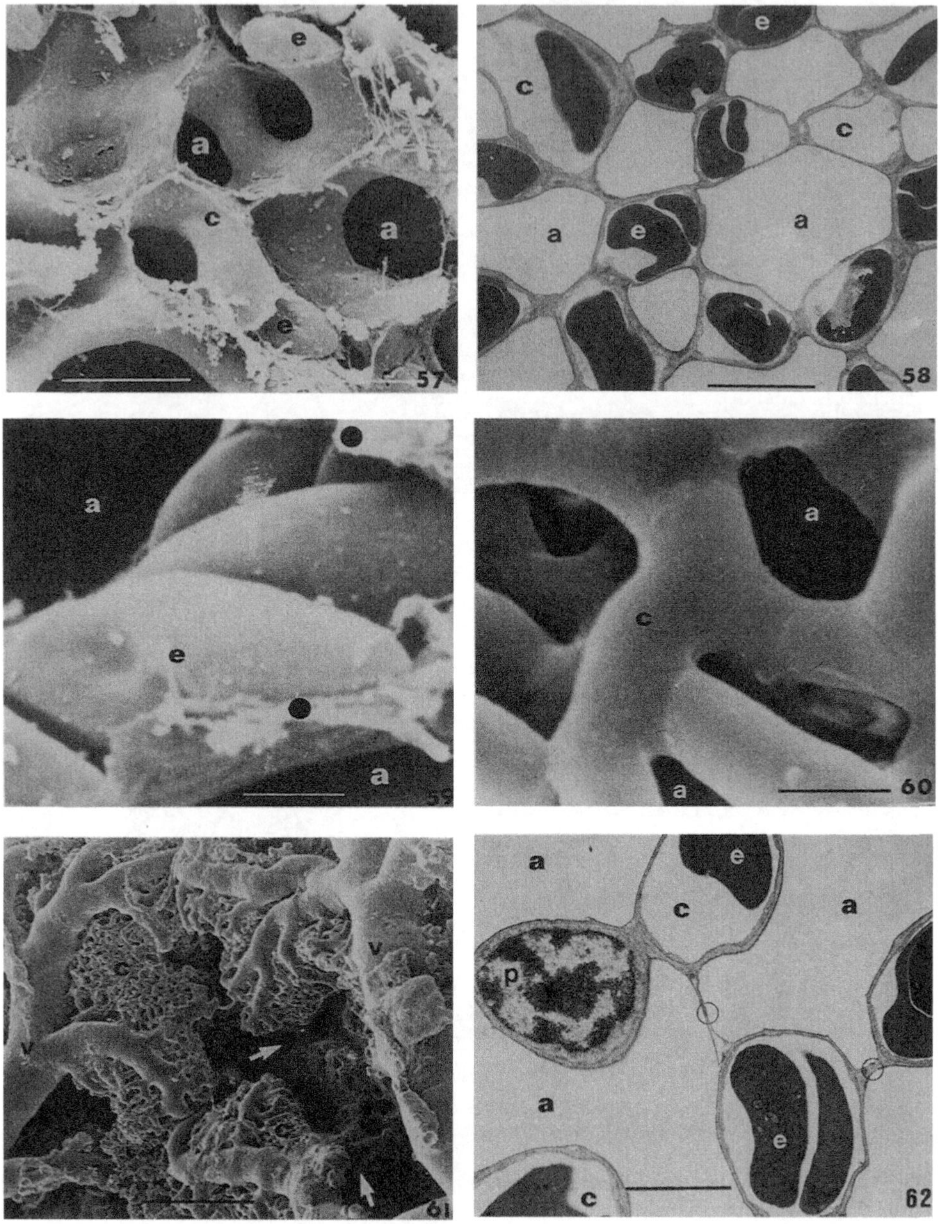

Figs. 57–62. Preparations of the lung of the domestic fowl, *Gallus gallus* var. *domesticus*. **57–60, 62** Components of the exchange tissue. *a*, air capillaries; *e*, red blood cells contained in the blood capillaries (*c*); *dots*, blood-gas (tissue) barrier separating red blood cells from the air capillaries; *p*, epithelial cell nucleus; *circle*, areas where air capillaries lie adjacent to each other. **61** Intraparabronchial blood vessels (*v*) that give rise to blood capillary systems (*c*) that interdigitate with the air capillaries arising from the parabronchi. *Arrows*, interatrial septa. *Scale bar*: **57**, 2 μm; **58**, 12 μm; **59**, 1 μm; **60**, 1.6 μm; **61**, 16 μm; **62**, 1.9 μm

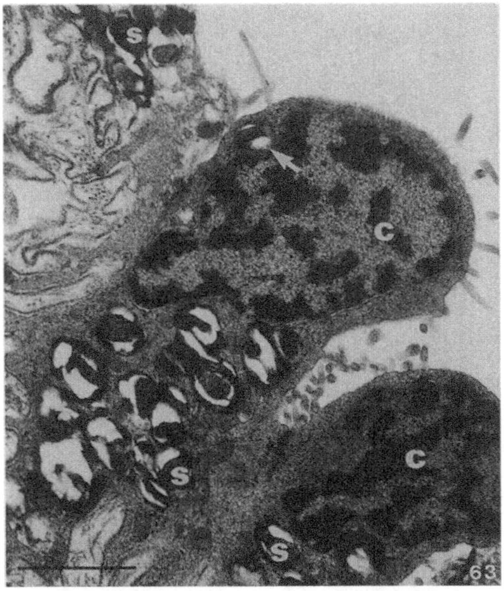

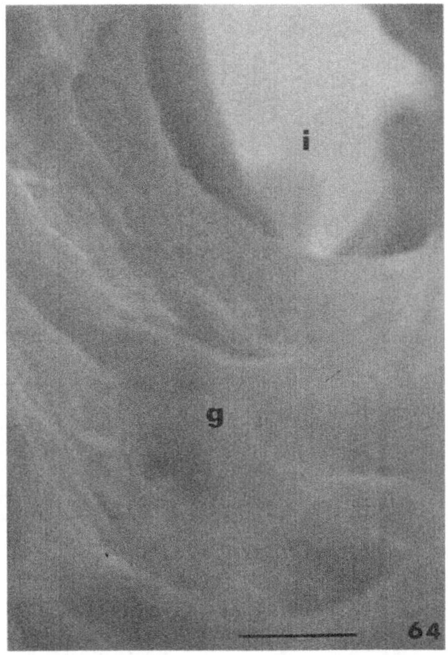

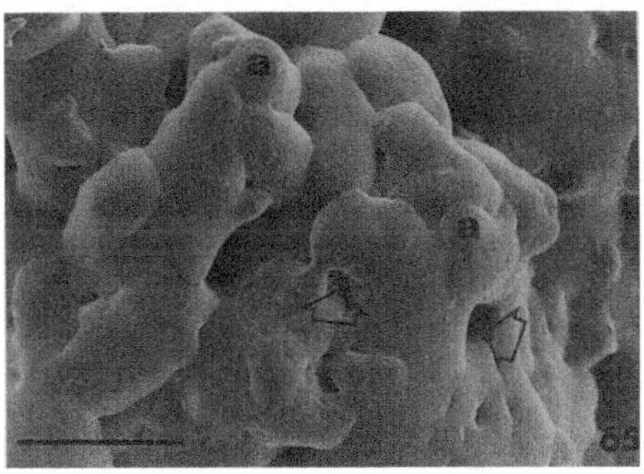

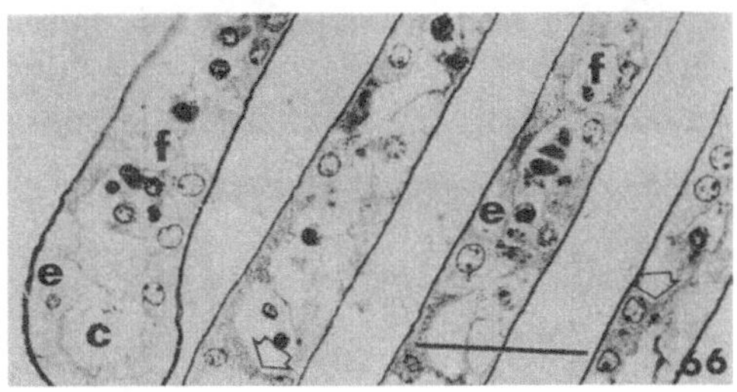

Fig. 63. Gas gland cells (*c*) lining the swimbladder of *Oreochromis alcalicus grahami*. The cells contain osmiophilic lamellated body-like structures (*s*) that appear to be secreted onto the free surface (*arrow*). *Scale bar*, 0.8 μm

Fig. 64. Floor of an infundibulum (*i*) of the lung of the rock dove, *Columba livia,* showing perikarya of type II cells (granular pneumocytes) (*g*). *Scale bar*, 1.3 μm

Fig. 65. Cast of the air capillaries (*a*) that interdigitate with blood capillaries (*arrows*) in the lung of the domestic fowl, *Gallus gallus* var. *domesticus*. *Scale bar*, 11 μm

Fig. 66. Gill filaments (*f*) of the gill arches of the fresh water crab, *Potamon niloticus*. *e*, epithelial cells that fuse at the middle of the gill filaments (*arrows*) subdividing it into vascular channels; *c*, marginal vascular channel. *Scale bar*, 78 μm

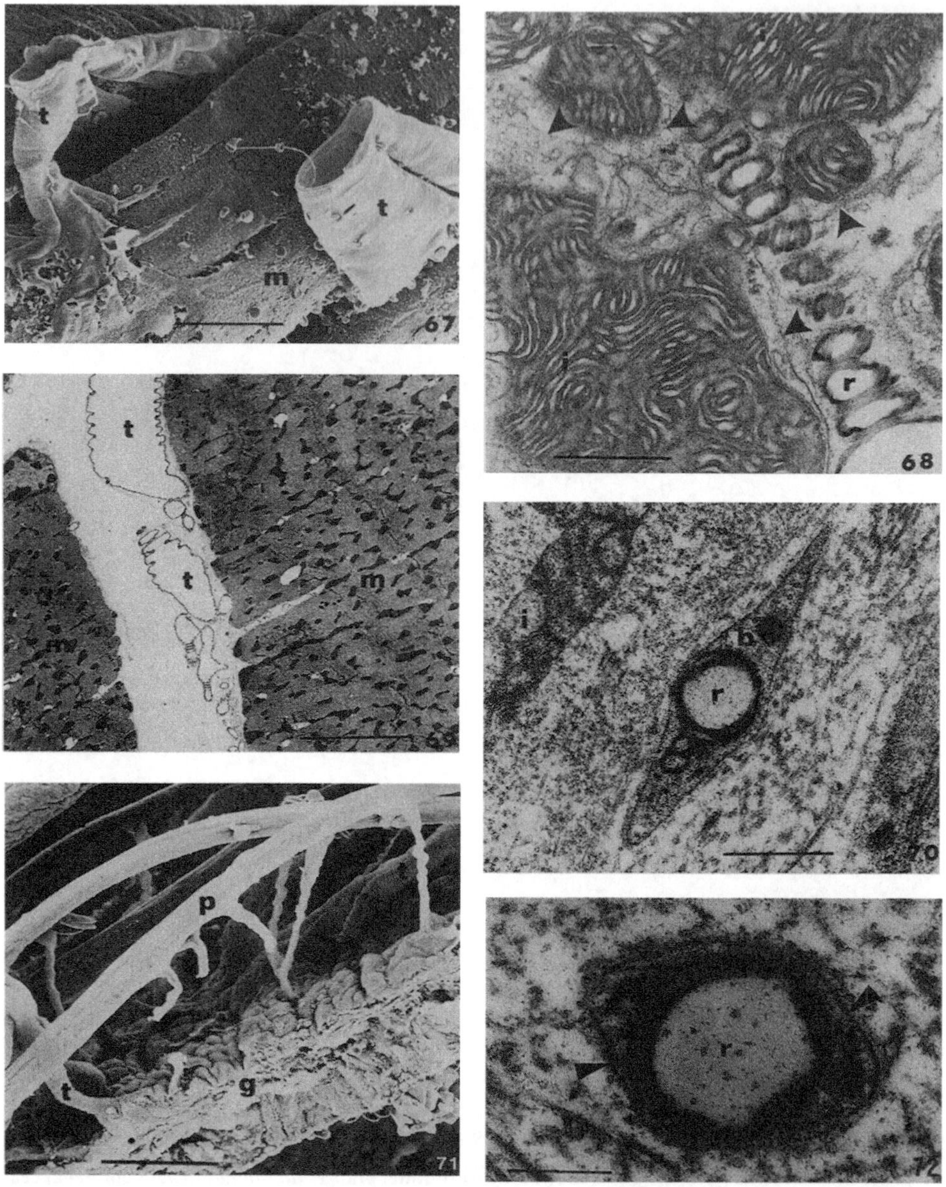

Figs. 67–72. Tracheal system of the grasshopper, *Chrotogonus senegalensis*. **67** Trachea (*t*) supplying air to the flight muscles (*m*). *Arrow*, a taenidium. *Scale bar*, 72 µm. **68** A tracheole (*r*) supplying air to mitochondria (*i*) in a flight muscle cell. The indentation of the tracheole into the cell is shown by *arrowheads* that mark the cell membrane. *Scale bar*, 0.6 µm. **69** Trachea (*t*) overlying a flight muscle (*m*). *Scale bar*, 6.5 µm. **70** A tracheole (*r*) in the tissue of the gastrointestinal system. *i*, a mitochondrion; *b*, a tracheoblast. *Scale bar*, 0.3 µm. **71** Trachea (*t*) supplying air to the wall of the gastrointestinal system (*g*). *p*, Malpighian tubules. *Scale bar*, 118 µm. **72** Cross section of a tracheole (*r*). *Arrowheads*, outer cell membrane of a tracheoblast. *Scale bar*, 0.1 µm

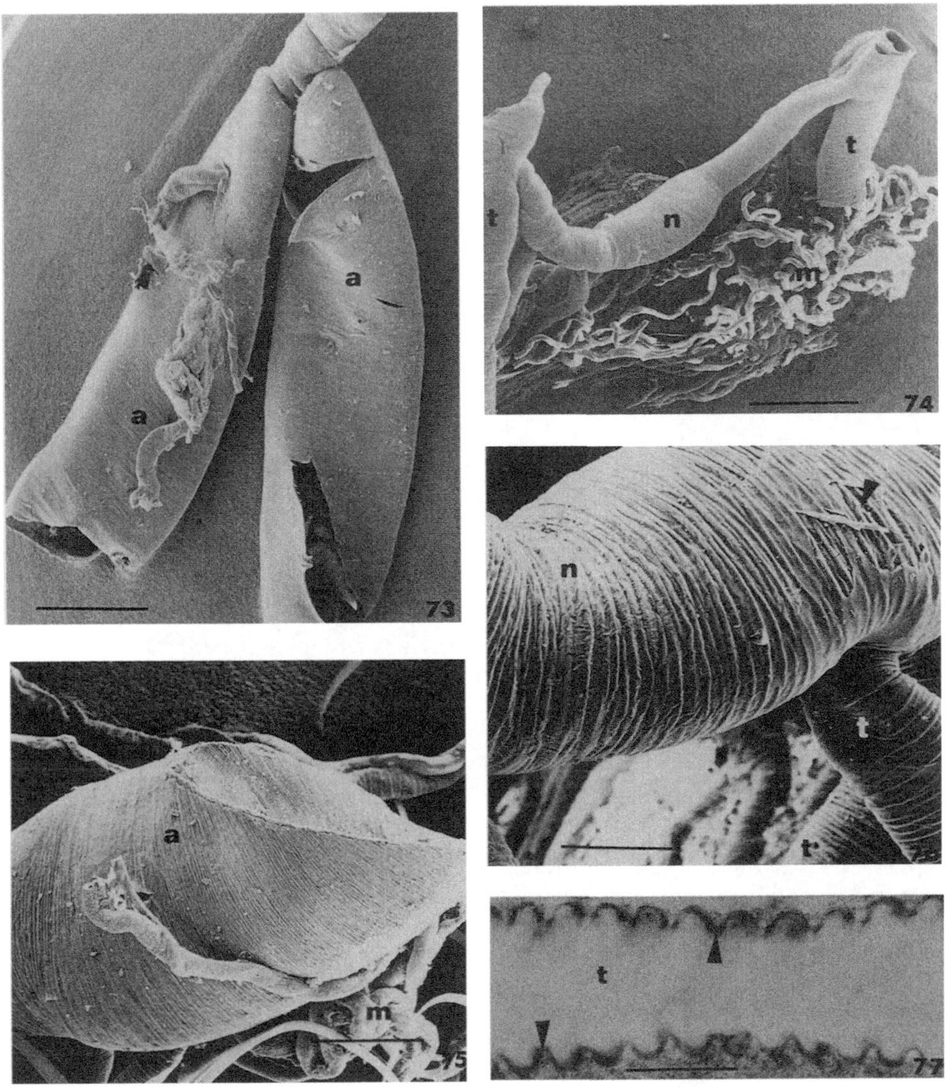

Figs. 73–77. Air sacs and trachea in the grasshopper, *Chrotogonous senegalensis*. *a*, air sacs; *t*, trachea; *m*, Malpighian tubules; *n*, longitudinal (anastomosing) trachea; *arrowhead*, taenidia. *Scale bar*: **73**, 0.8 mm; **74**, 1.2 mm; **75**, 119 μm; **76**, 3 mm; 77, 0.4 μm

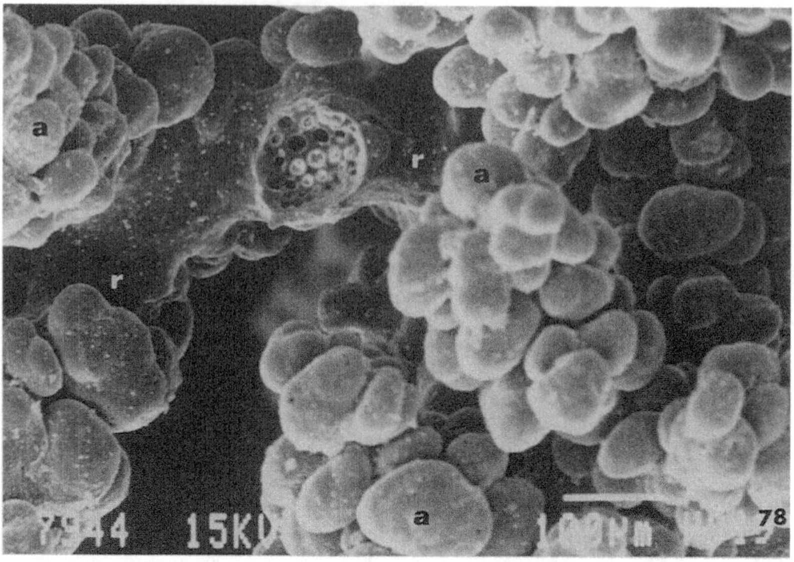

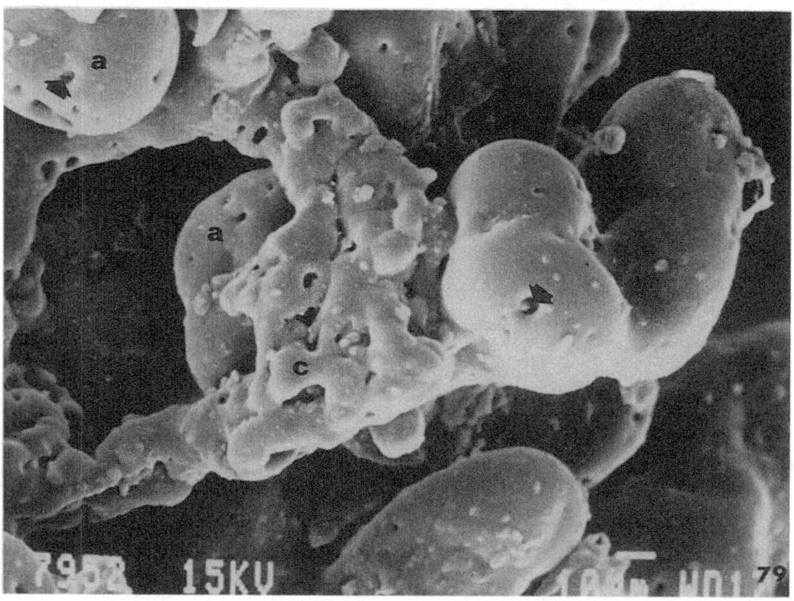

Figs. 78, 79. Lung of the domestic pig, *Sus scrofa*, showing alveoli (*a*), respiratory bronchioles (*r*) and blood capillaries (*c*) encapsulating alveoli. *Arrows*, interalveolar pores. *Scale bar*: **78**, 100 μm; **79**, 10 μm

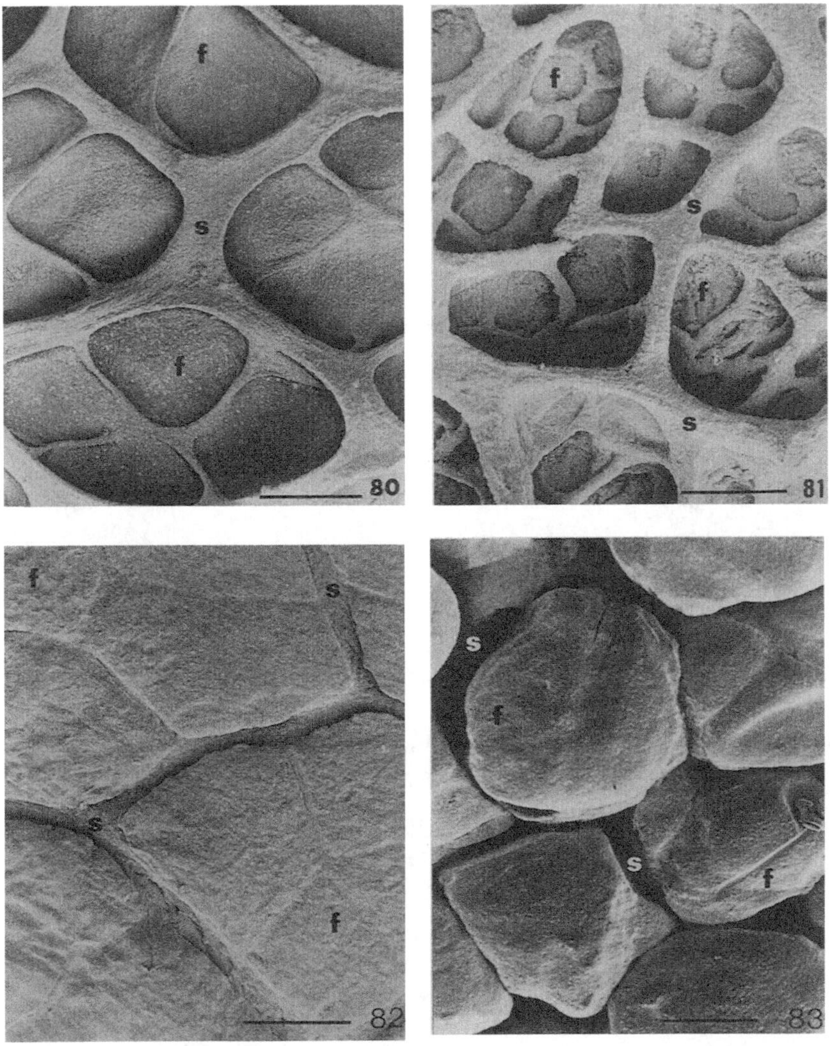

Figs. 80–83. Compartmentalization of the exchange tissue of the lungs of the monitor lizard, *Varanus exanthematicus* (**80** and **82**), and the pancake tortoise, *Malacochersus tornieri* (**81** and **83**). *f*, faveoli; *s*, septa. *Scale bar*: **80**, 0.3 mm; **81**, 0.4 mm; 82, 0.13 mm; **83**, 0.3 mm

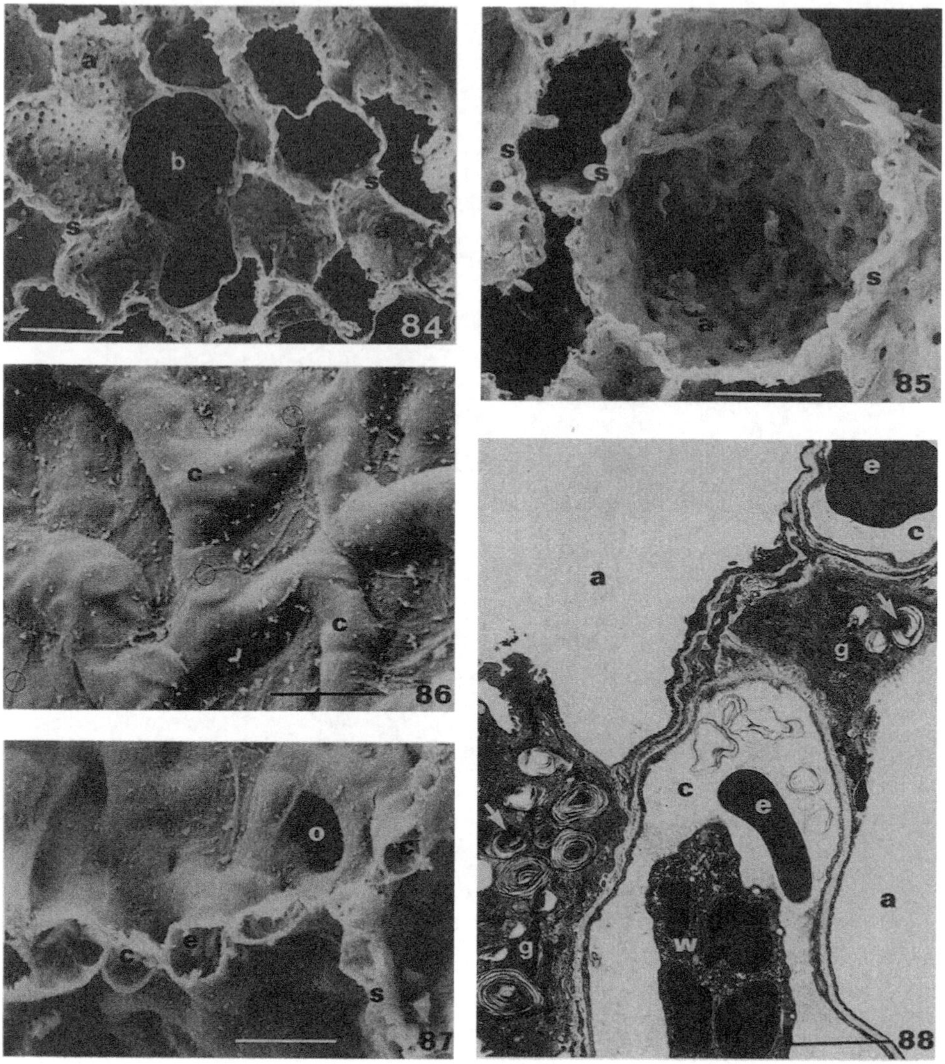

Figs. 84–88. Lung of the lesser bushbaby, *Galago senegalensis,* showing alveoli (*a*), a respiratory bronchiole (*b*), interalveolar septum (*s*), blood capillaries (*c*), interalveolar pores (*o*), red blood cells (*e*), type II cells (*g*), osmiophilic lamellated bodies (*arrows*), a leukocyte (*w*) and intercellular junctions of the type I cells (*circles*). *Scale bar:* **84,** 95 μm; **85,** 30 μm; **86,** 8 μm; **87,** 0.7 μm; **88,** 1.4 μm

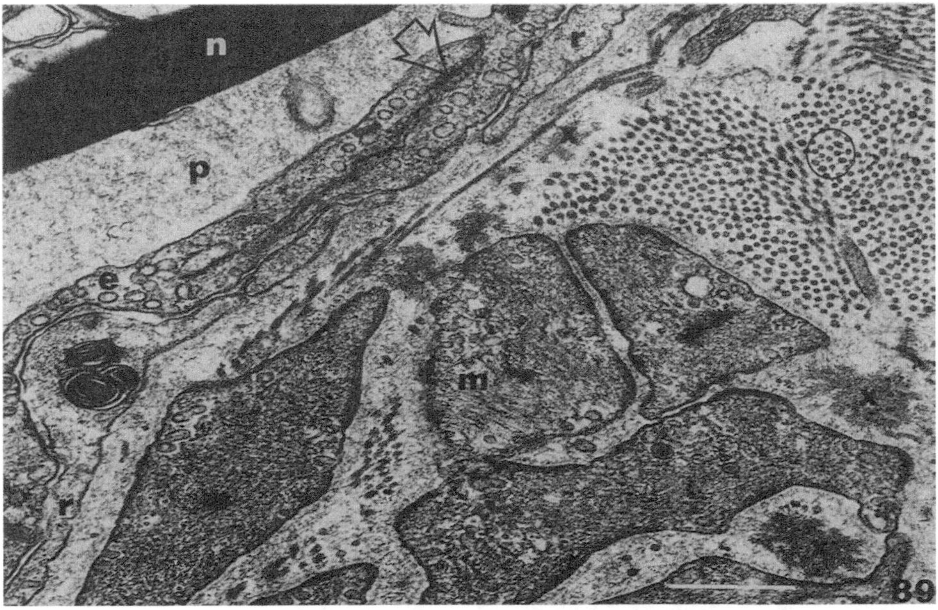

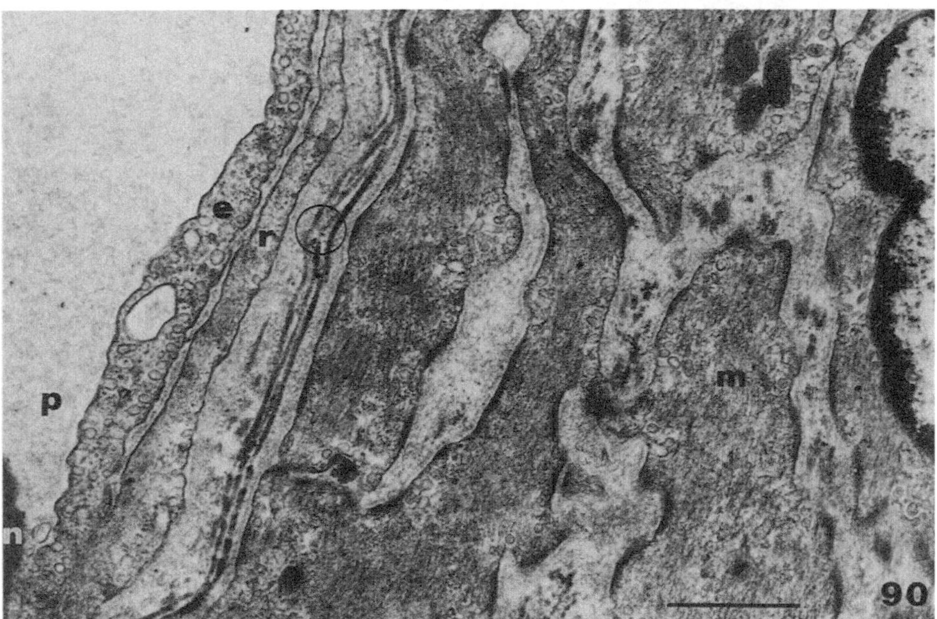

Figs. 89, 90. Smooth muscles (*m*), collagen (*circles*) and elastic tissue (*x*) in the lungs of the black mamba, *Dendroapis polylepis* (**89**), and the tree frog, *Chiromantis petersi* (**90**). *e*, endothelial cell; *arrow*, endothelial cell junction; *p*, plasma layer; *r*, pericyte; *n*, erythrocyte in a blood capillary. *Scale bar*: **89**, 0.9 µm; **90**, 0.7 µm

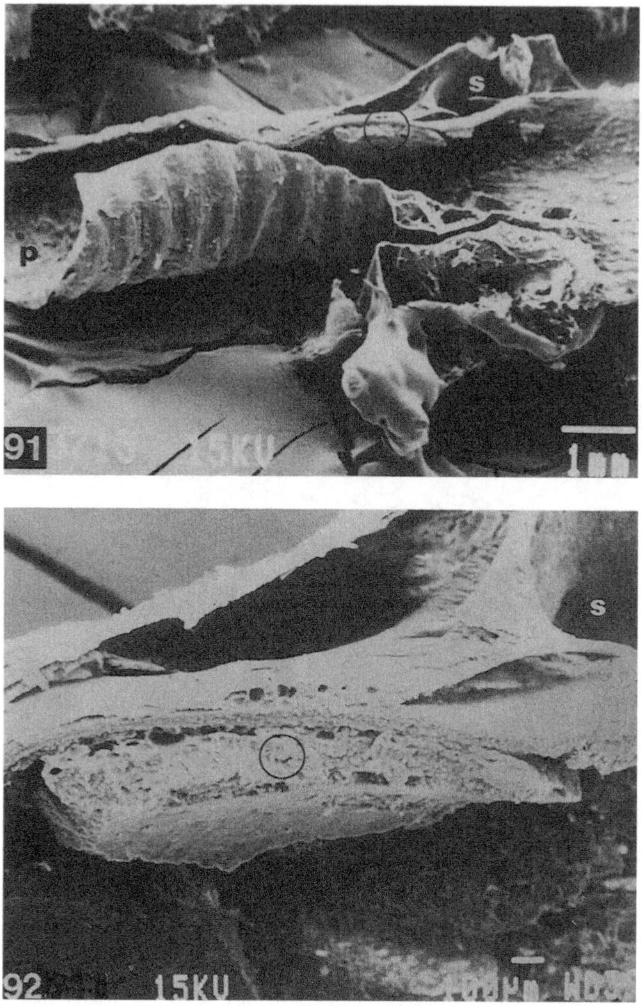

Figs. 91, 92. Extrapulmonary primary bronchus (*p*) of the lung of the domestic fowl, *Gallus gallus* var. *domesticus*. Side view showing an epithelial swelling (*circle*), the segmentum accerelans, just before the origin of the first medioventral secondary bronchus (*s*). *Scale bar*: **91**, 1 μm; **92**, 100 μm

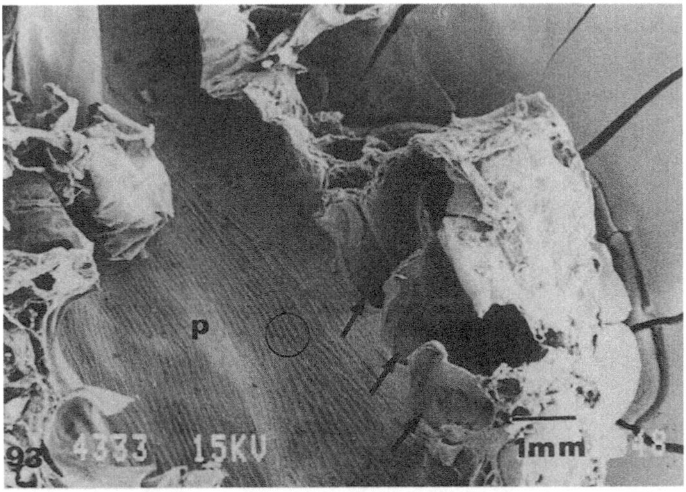

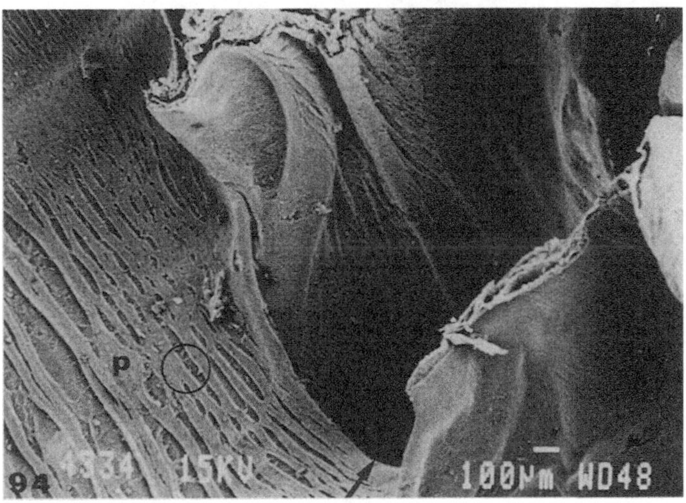

Figs. 93, 94. The lung of the domestic fowl, *Gallus gallus* var. *domesticus*, with the medioventral secondary bronchi (*arrows*) arising from the extrapulmonary primary bronchus (*p*). The *circles* show conspicuous epithelial tracts. *Scale bar*: **93**, 1 mm; **94**, 100 μm

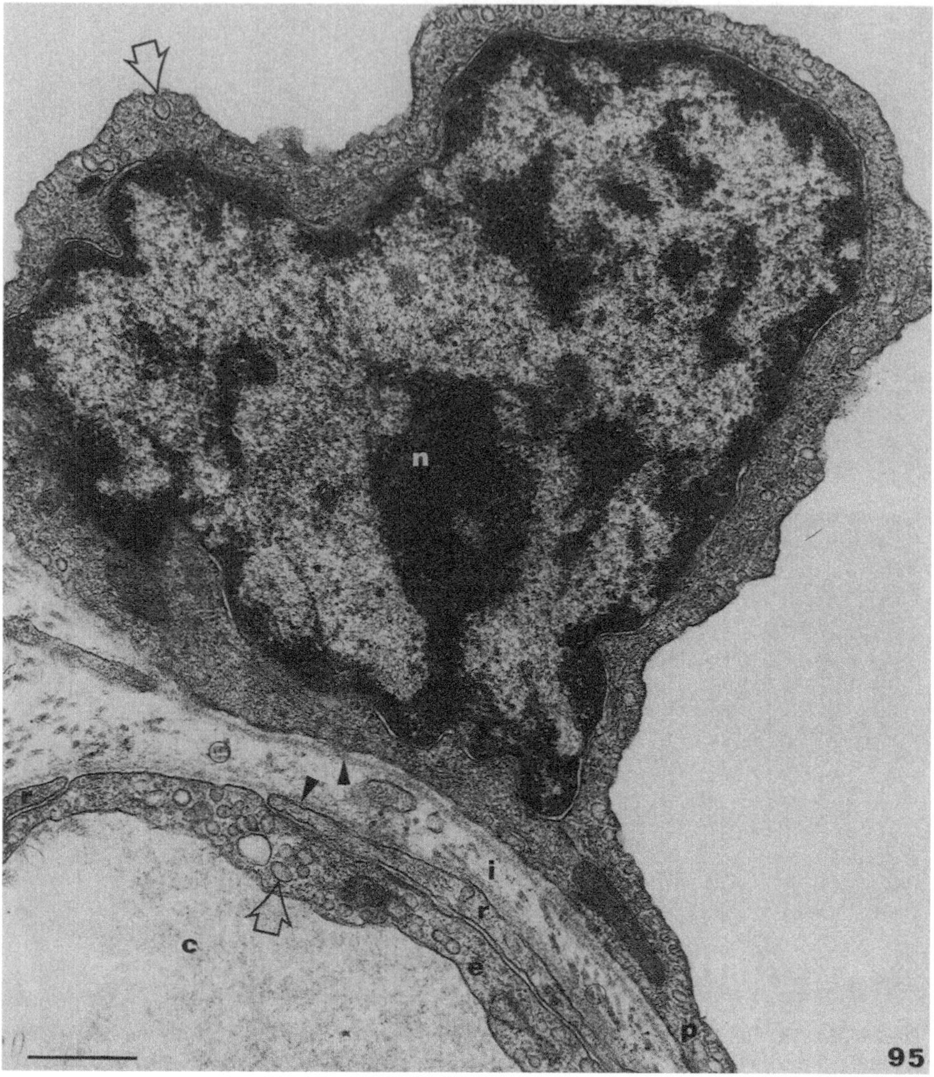

Fig. 95. Perikaryon of a type I cell (*n*) of the lung of the tree frog, *Chiromantis petersi*, lying on a blood capillary (*c*). The cell gives rise to thin cytoplasmic extensions (*p*) that cover much of the respiratory surface. *Arrowheads*, basal membranes of the epithelial and endothelial cells (*e*) enclosing an interstitial space (*i*) that contains collagen fibres (*circles*). *Arrows*, micropinocytotic vesicles; *r*, pericytes. *Scale bar*, 0.2 μm

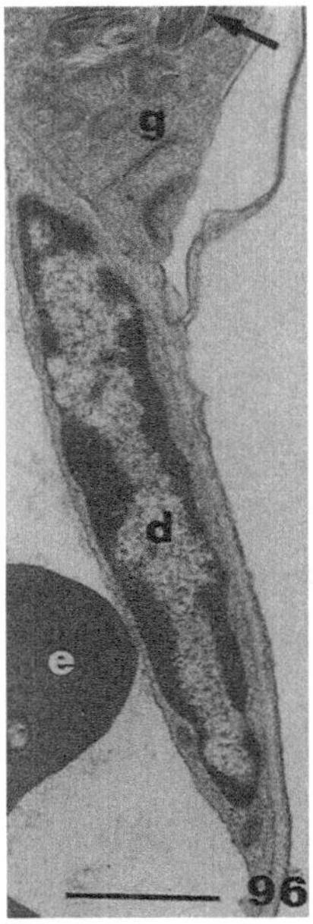

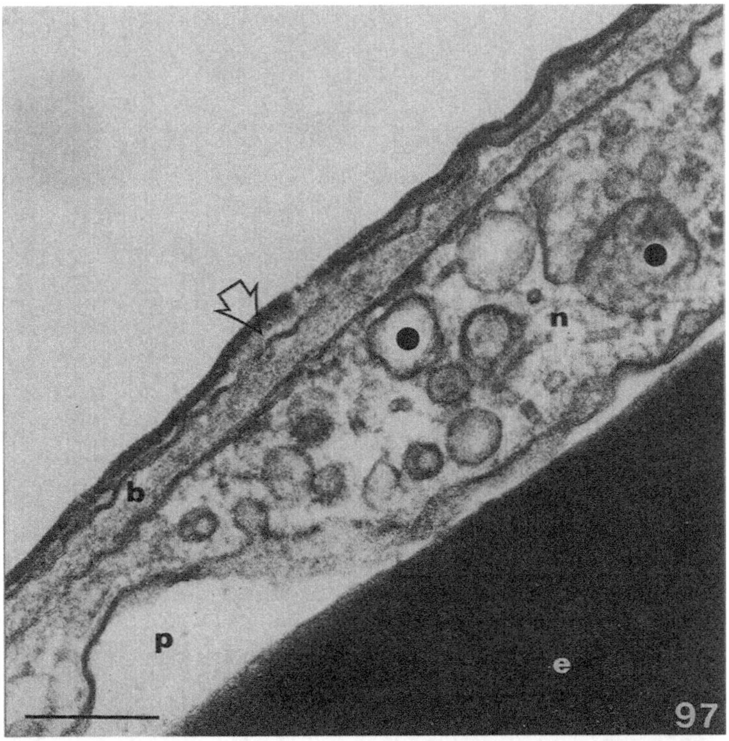

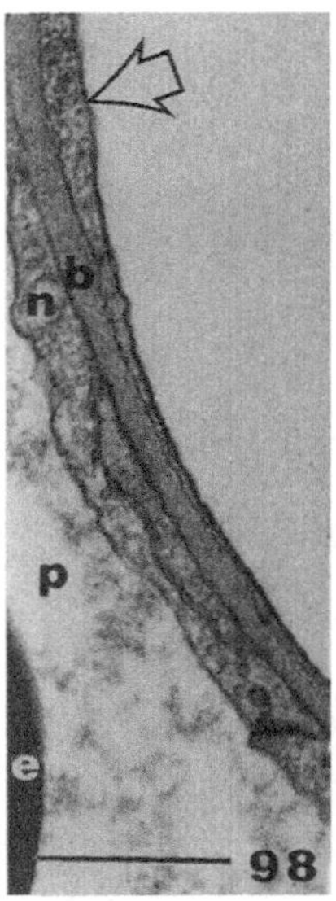

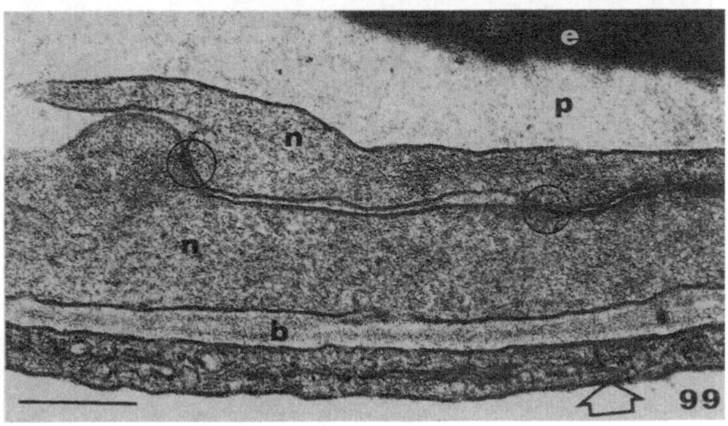

Fig. 96. Blood-gas (tissue) barrier of the lung of the emu, *Dromiceius novaehollandiae,* showing a type II pneumocyte (*g*). *Arrow,* osmiophilic lamellated body; *d,* endothelial cell; *e,* red blood cell. *Scale bar,* 1.8 μm

Figs. 97, 98. The blood-gas (tissue) barrier of the lung of the black-headed gull, *Larus ridibundus* showing the epithelial cell (*arrows*); common basement membrane (*b*); endothelial cell (*n*); plasma layer (*p*), red blood cell (*e*) and micropinocytotic vesicles (*dots*). *Scale bar:* **97,** 0.15 μm; **98,** 0.4 μm

Fig. 99. Blood-gas (tissue) barrier of the lung of the black mamba, *Dendroaspis polylepis* showing capillary endothelial cells (*n*) fusing at interrupted tight junctions (*circles*). *b,* common basement membrane; *arrow,* epithelial cell; *e,* red blood cell; *p,* plasma layer. *Scale bar,* 0.3 μm

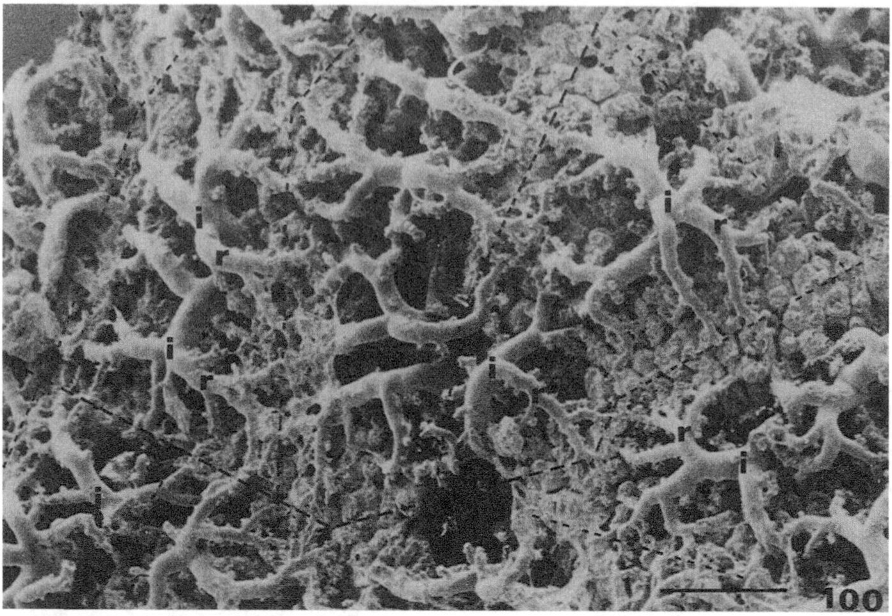

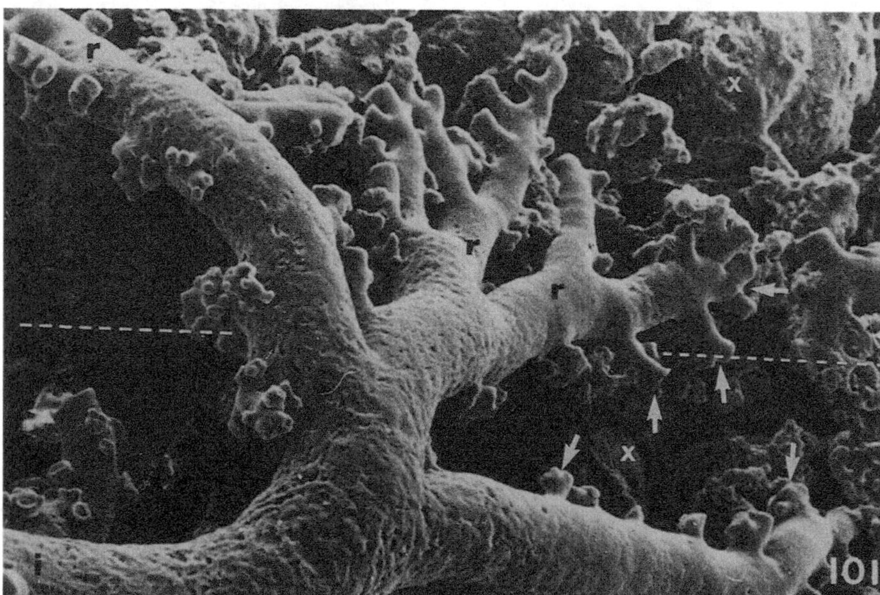

Figs. 100, 101. Double latex cast preparation (latex rubber injected into the airways and into the arterial vasculature) of the lung of the domestic fowl, *Gallus gallus* var. *domesticus*, to show the cross-current design between the blood flow and the parabronchial air flow. The venous blood in the intrapulmonary arterioles (*r*) flows in a perpendicular direction in relation to that of air in the parabronchial lumen (directions shown by *dashed lines*). *i*, interparabronchial arteries; *x*, atria arising from parabronchi; *arrows*, areas where blood capillaries and air capillaries contact each other. *Scale bar*: **100**, 0.2 mm; **101**, 9 µm

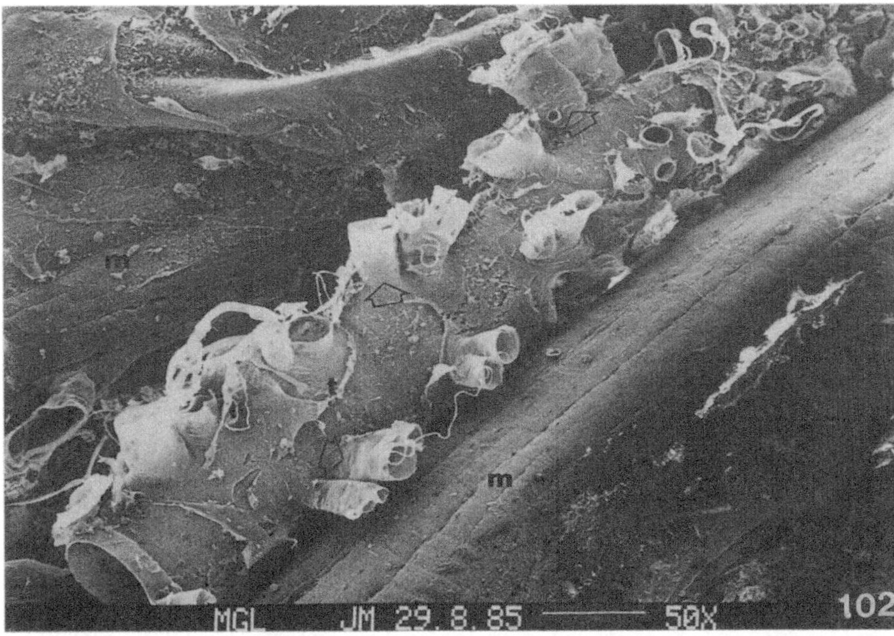

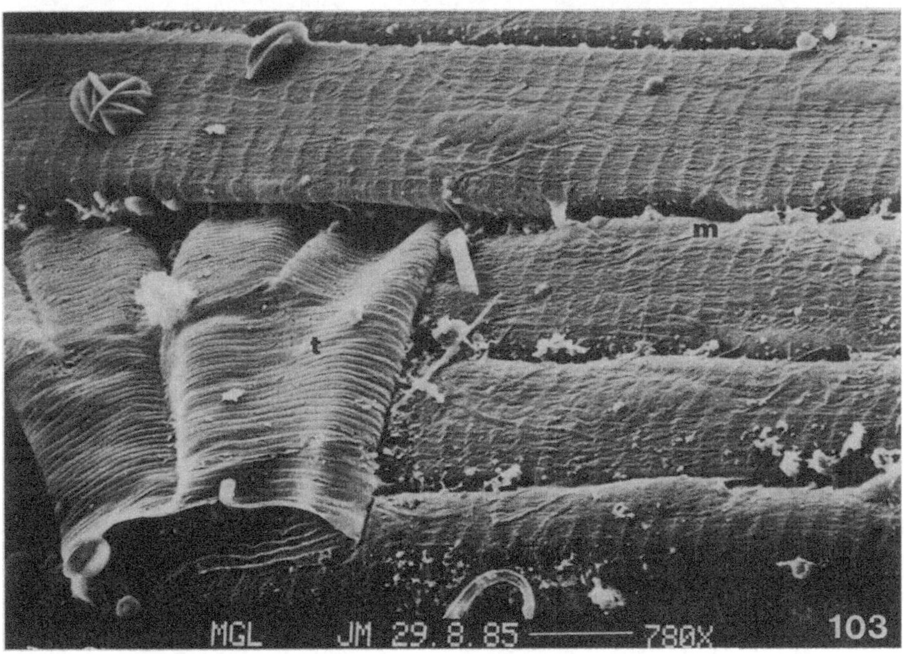

Figs. 102, 103. Tracheal air supply to the flight muscles of the grasshopper, *Chrotogonus senegalensis*. *t*, trachea; *m*, flight muscle; *arrows*, tracheal branches. *Scale bar*: **102,** 0.5 mm; **103,** 26 μm

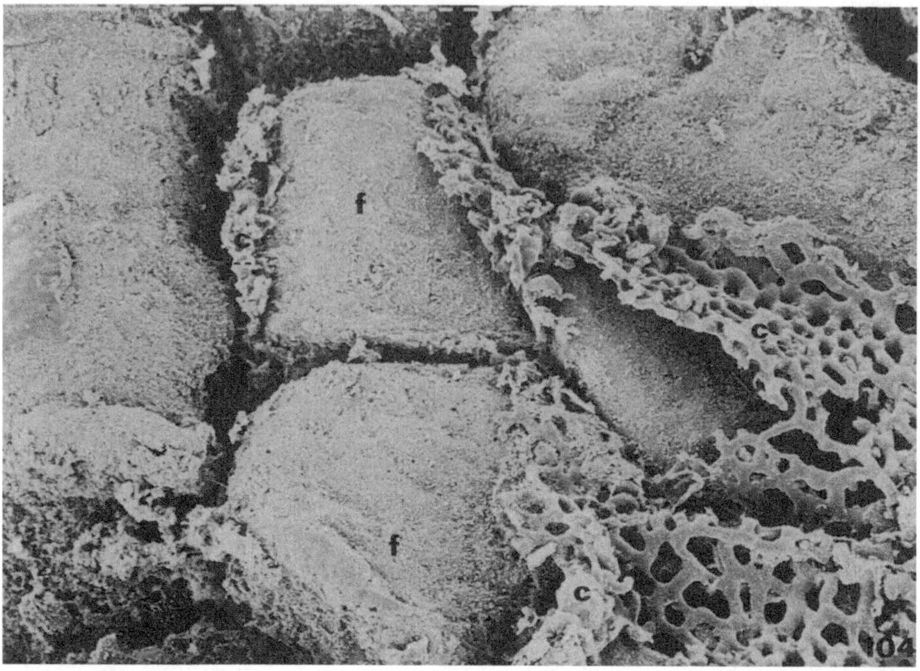

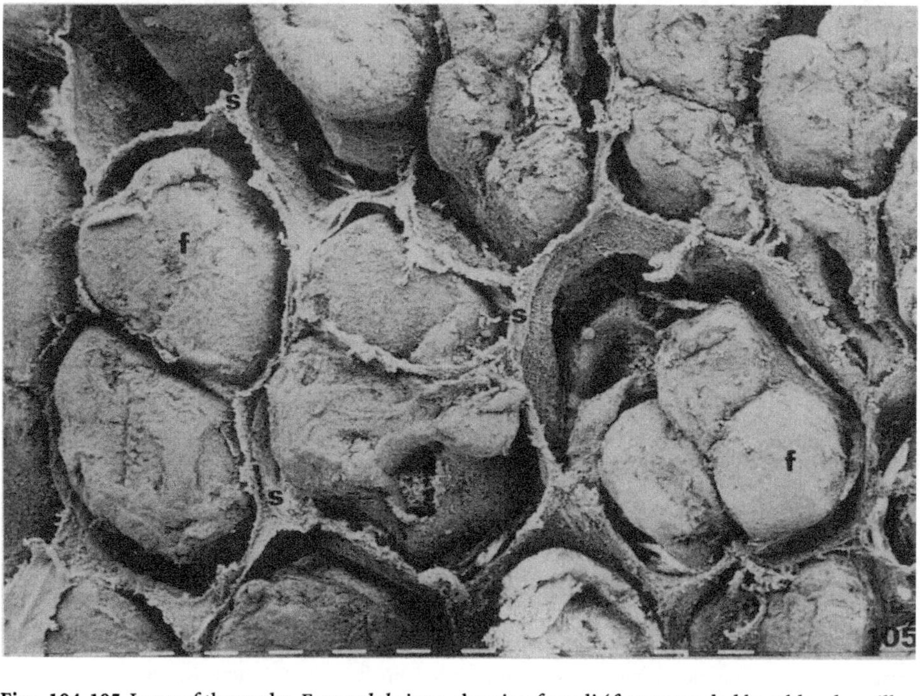

Figs. 104, 105. Lung of the snake, *Eryx colubrinus*, showing faveoli (*f*) surrounded by a blood capillary network (*c*) contained in interfaveolar septa (*s*). *Scale bar*: **104**, 0.8 µm; **105**, 0.02 mm

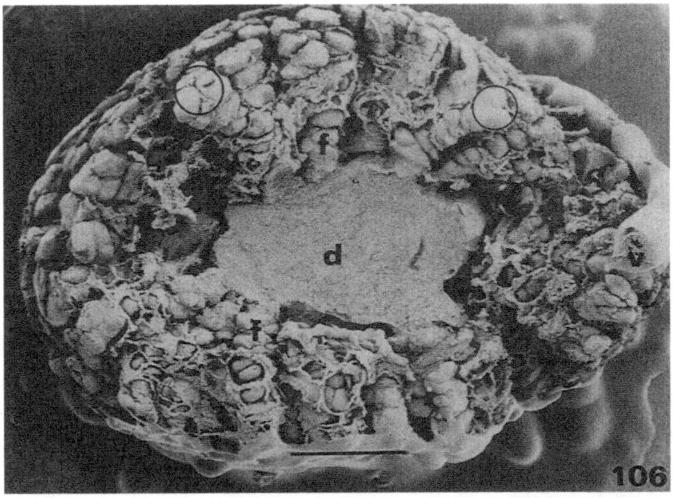

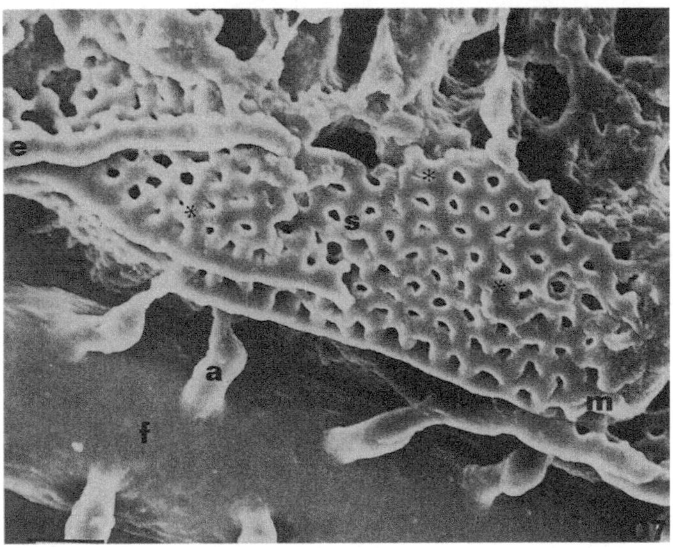

Fig. 106. Lung of the snake, *Eryx colubrinus,* showing a central air duct (*d*) with faveoli (*f*) radiating outwards. The faveoli divide into about three air cells (*circles*). *v*, pulmonary artery. *Scale bar*, 27 µm

Fig. 107. Gills of *Oreochromis alcalicus grahami* showing a gill ilament vessel (*f*) and a secondary lamella (*s*). *Asterisks*, vascular channels of the secondary lamellae that are separated by pillar cells (*holes*). *m*, marginal channel; *a*, afferent arteriole; *e*, efferent arteriole. *Scale bar*, 60 µm

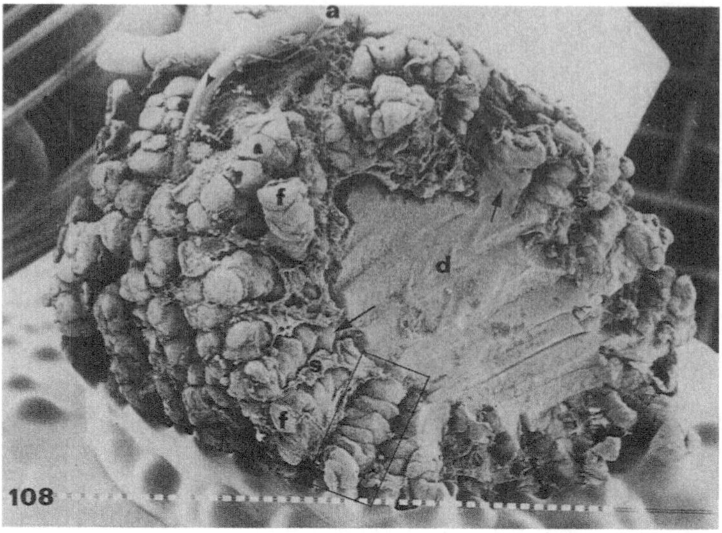

Figs. 108, 109. The snake lung, *Eryx colubrinus,* showing a central air duct (*d*) and faveoli (*f*) radiating outwards (*arrows*). A pulmonary artery (*a*) gives rise to an intrapulmonary branch (*arrowhead*) that supplies blood to capillary system of the interfaveolar septa (*s*). **109** An enlargement of the faveolus (enclosed in **108**). *Dots,* constrictions of the interfaveolar walls. The *inset* shows faveoli (*dashed lines*) on a fixed tissue preparation. *Scale bar:* **108,** 27 μm; **109,** 17 μm; *inset:* 1 mm

Fig. 110. Alveolar septum of the lung of the naked mole rat, *Heterocephalus glaber,* showing an alveolar space (*a*) and a sheet of blood capillaries (*c*). *Scale bar,* 12 μm

Fig. 111. Longitudinal view of a gill filament (*f*) of *Oreochromis alcalicus grahami* showing the secondary lamellae (*s*) arising from opposite sides a gill filament. *Asterisks,* vascular channels; *e,* epithelial lining that has pulled away. *Scale bar,* 2 μm

Fig. 112. Vascularized surface of the swimbladder of *Oreochromis alcalicus grahami* showing blood vessels (*v*). *Scale bar,* 13 μm

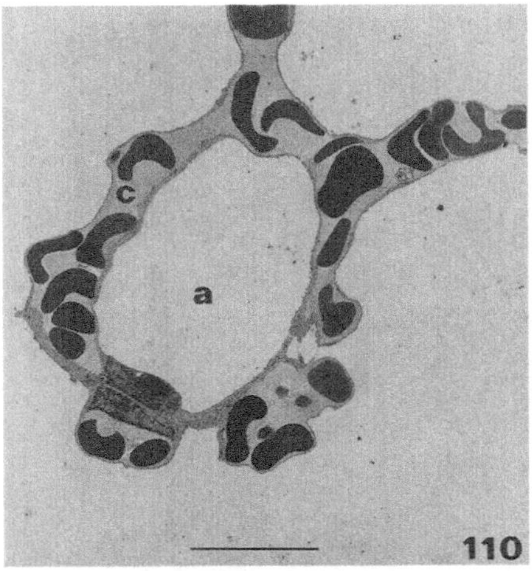

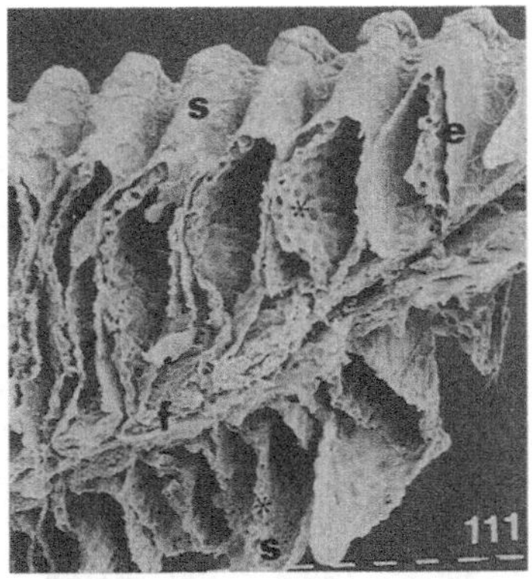

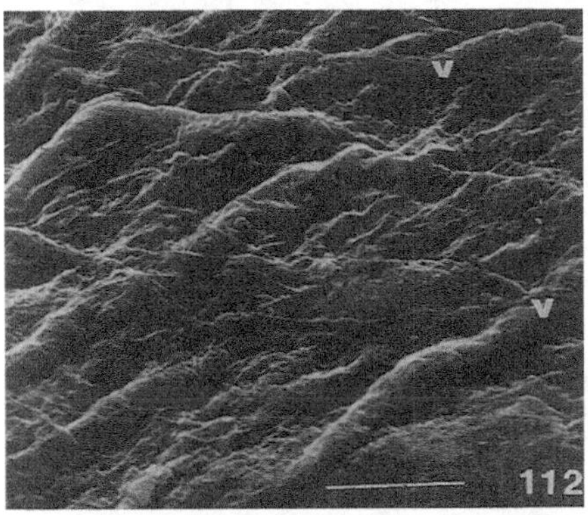

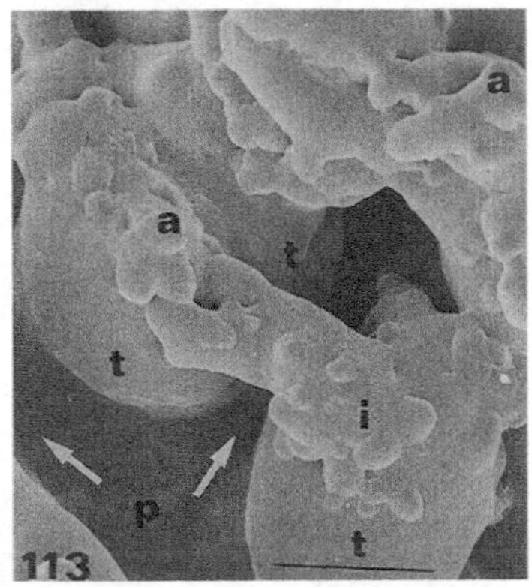

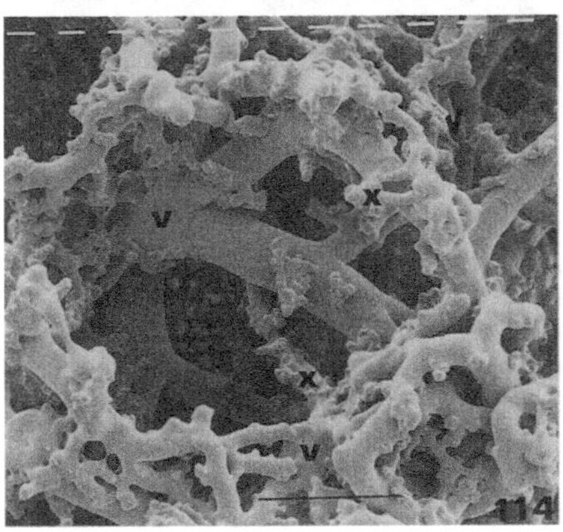

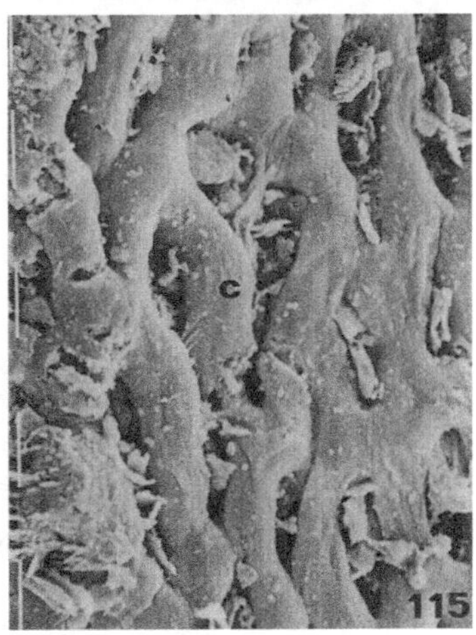

Fig. 113. Parabronchus (*p*) of the lung of the domestic fowl, *Gallus gallus* var. *domesticus,* showing atria (*t*) giving rise to infundibulae (*i*) and the terminal gas exchange components, the air capillaries (*a*). *Arrows,* interatrial septa. *Scale bar,* 0.1 mm

Fig. 114. Parabronchus of the lung of the domestic fowl, *Gallus gallus* var. *domesticus,* showing the interparabronchial (*v*) and intraparabronchial vessels (*x*). *Scale bar,* 0.1 mm

Fig. 115. Anastomosing blood capillaries (*c*) on the skin of the tree frog, *Chiromantis petersi. Scale bar,* 27 μm

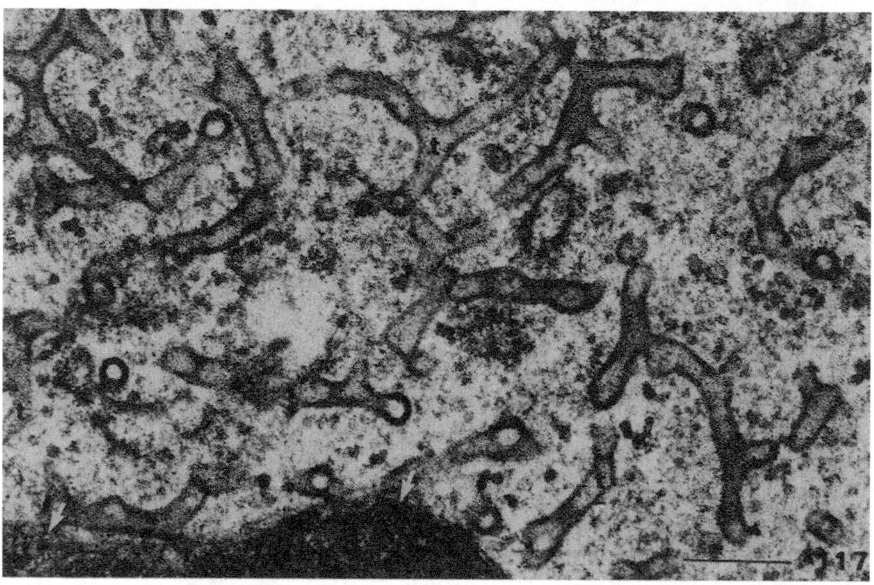

Fig. 116. Surface of the suprabranchial chamber membrane of the African catfish, *Clarias mossambicus*, showing tracts of blood capillaries (*c*) separated by clear sites (*arrowheads*) where goblet (mucus secreting) cells are located. *Scale bar*, 14 μm

Fig. 117. Profuse microtubular network in the chloride cells of the gills of *Oreochromis niloticus*. *arrow*, mitochondrion. *Scale bar*, 0.1 μm

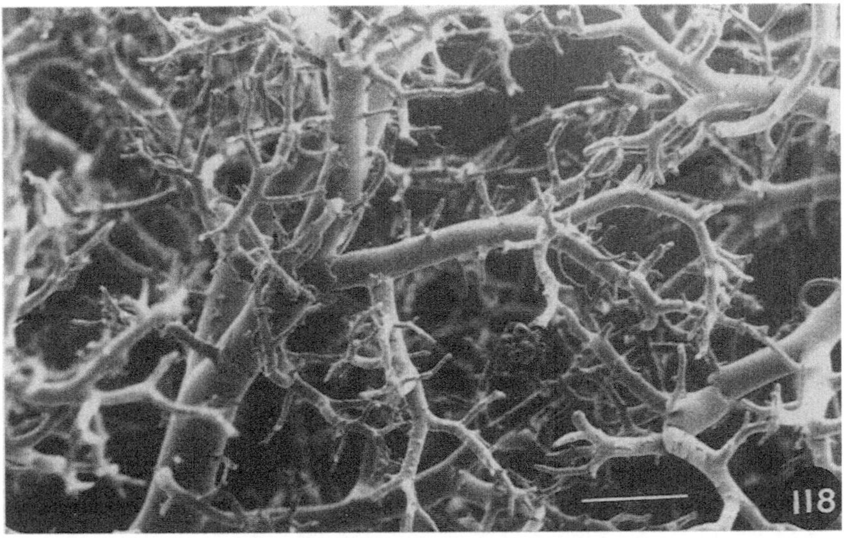

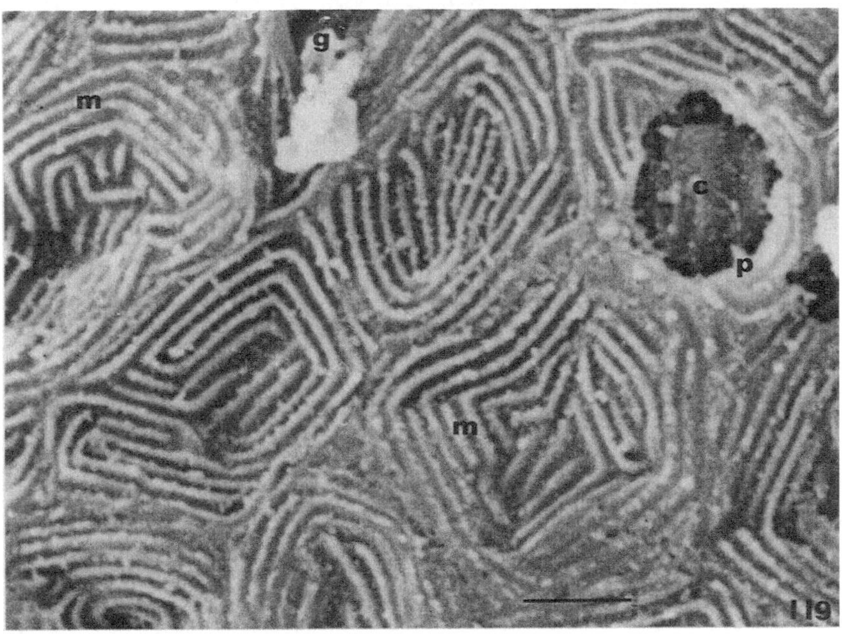

Fig. 118. The fractal (tree-like) nature of the vascular system of pulmonary arterial system of the lung of the domestic pig, *Sus scrofa*. *Scale bar*, 0.7 mm

Fig. 119. Surface of the gill arch of *Oreochromis alcalicus grahami* showing chloride cells (*c*) surrounded by pavement cells (*p*). *g*, a mucus plug secreted by a goblet cell; *m*, microridges. *Scale bar*, 3 μm

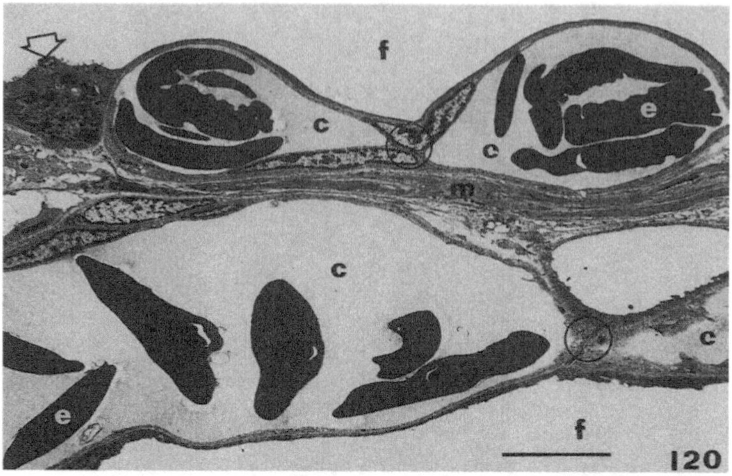

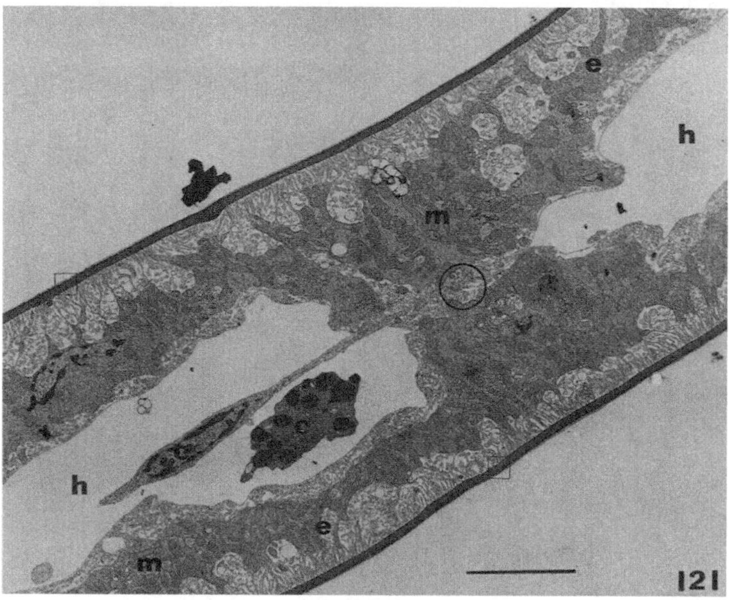

Fig. 120. Pulmonary septum of the lung of the pancake tortoise, *Malacochersus tornieri*, showing blood capillaries (*c*) exposed to air in the faveoli (*f*) only on one side. The blood capillaries are separated by fusion of endothelial cells (*circles*). The septum contains smooth muscle elements (*m*). *e*, red blood cells; *arrow*, a pneumocyte with osmiophilic lamellated bodies. *Scale bar*, 3 μm

Fig. 121. Gill filament of a fresh water crab, *Potamon niloticus*, showing epithelial cells (*e*) fusing at the middle (*circle*) and separating the filament into haemolymphatic channels (*h*). *m*, mitochondrial aggregations; *squares*, cuticular lining of the gill filaments; *c*, haemocytes of which one has been entrapped between fusing epithelial cells. *Scale bar*, 7 μm

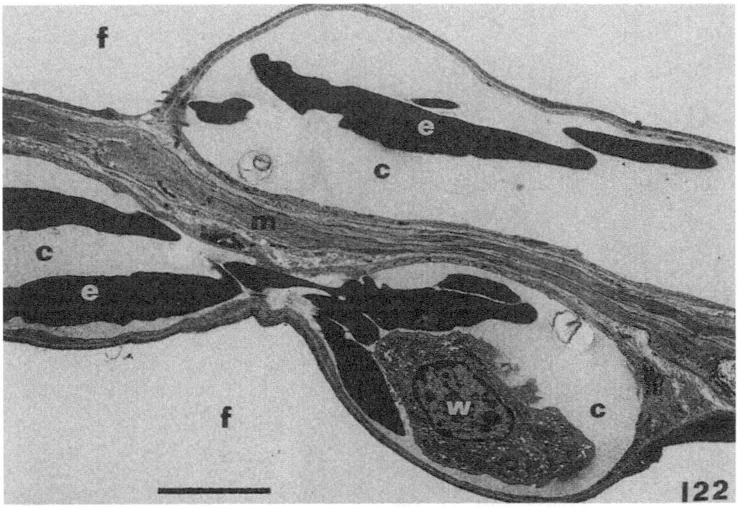

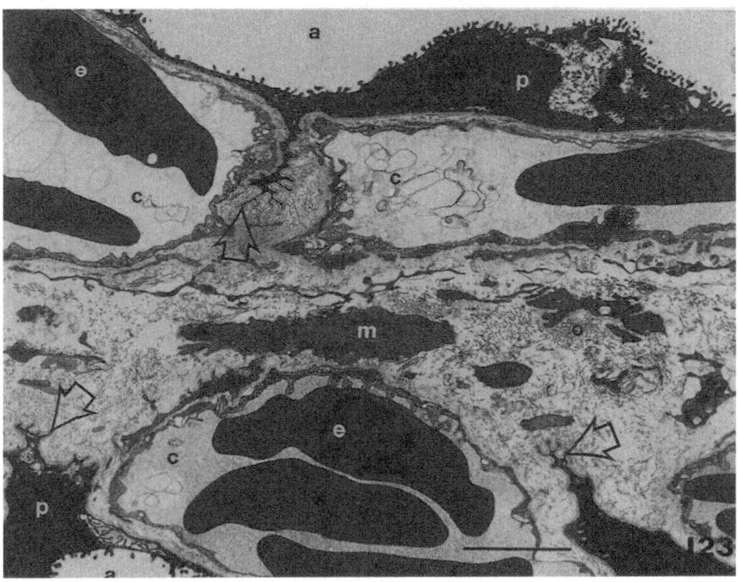

Fig. 122. Septum of the lung of the snake, *Dendroaspis polylepis,* showing blood capillaries (*c*) exposed to air in the faveoli (*f*) only on one side. The septum contains smooth muscle elements (*m*). *e*, red blood cells; *w*, leukocyte. *Scale bar*, 9 µm

Fig. 123. Septum of the lung of the caecilian, *Boulengerula taitanus,* showing blood capillaries (*c*) exposed to air (*a*) only on one side. Epithelial cell involutions (*arrows*) separate the blood capillaries. *e*, red blood cells; *p*, epithelial cells; *white arrowhead*, an osmiophilic lamellated body in an undifferentiated pneumocyte; *m*, smooth muscle; *o*, collagen. *Scale bar*, 3 µm

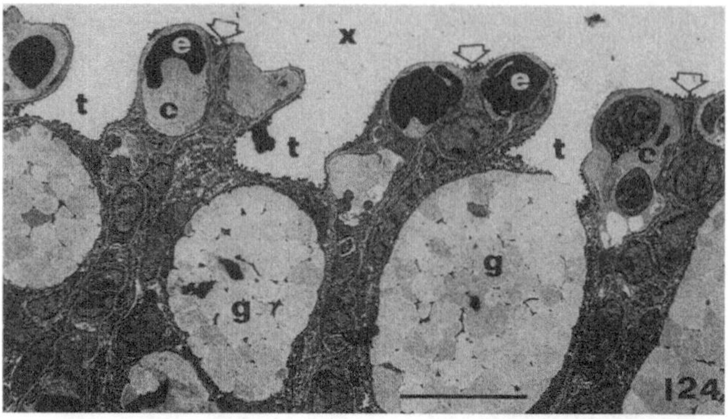

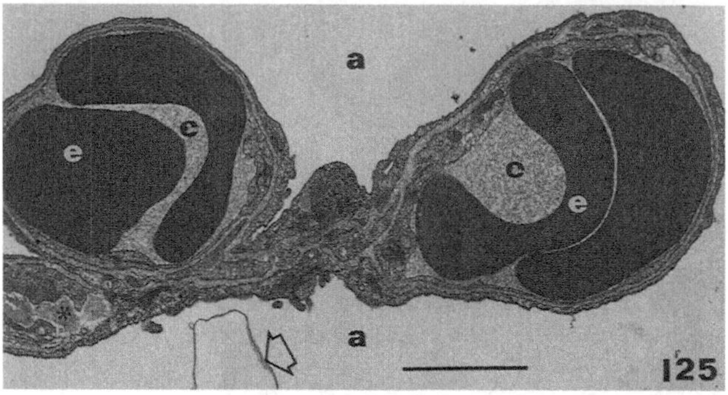

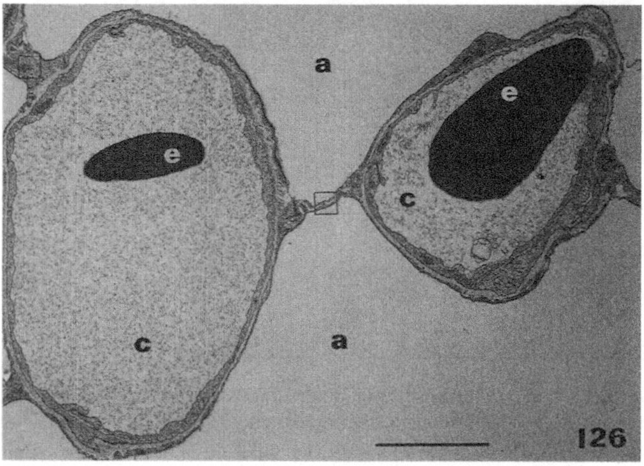

Fig. 124. A cross sectional view of the blood capillaries (*c*) that line the labyrinthine organ of the African catfish, *Clarias mossambicus*. *e*, red blood cells; *g*, mucus secretory (goblet) cells contained in tracts (*t*) that separate the blood capillaries. *Arrows*, epithelial cells separating blood capillaries; *x*, air space. *Scale bar*, 27 µm

Fig. 125. Blood capillaries (*c*) of the lung of the bat, *Epomophorus wahlbergi*, containing red blood cells (*e*). *a*, alveoli; *asterisk*, elastic tissue; *arrow*, surfactant discharged onto the alveolar surface. *Scale bar*, 2 µm

Fig. 126. Blood capillaries (*c*) of the lung of the domestic fowl, *Gallus gallus* var. *domesticus*, containing red blood cells (*e*). *a*, air capillaries; *squares*, areas where air capillaries lie in direct contact and where a common basement membrane is lacking. *Scale bar*, 3 µm

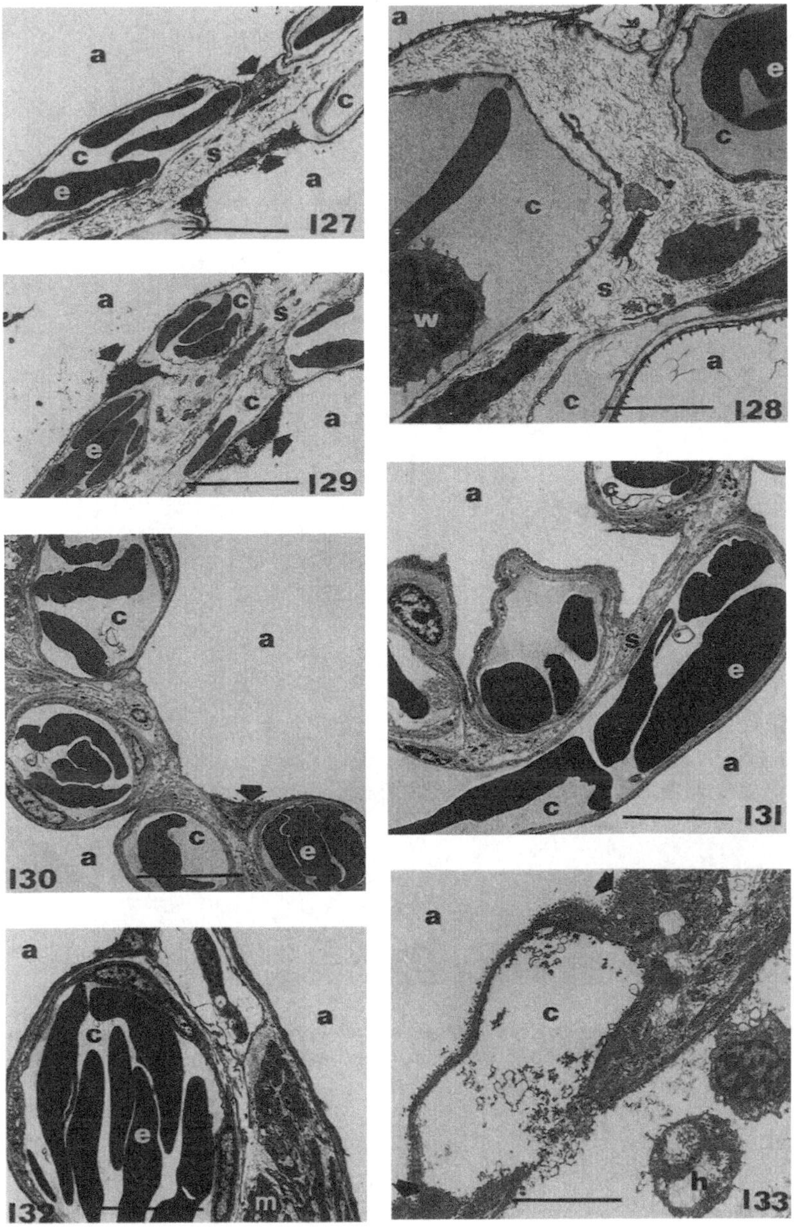

Figs. 127–133. Exposure of pulmonary capillary blood to air in lungs of different animals. **127** Lung of the tree frog, *Chiromantis petersi*; **128** Lung of the caecilian, *Boulengerula taitanus*; **129** Lung of the African lungfish, *Protopterus aethiopicus*; **130** Lung of the snake, *Dendroaspis polylepis*; **131** Lung of the tree frog, *Chiromantis petersi*; **132** Lung of the snake, *Dendroaspis polylepis*; **133** Lung of a slug, *Trichotoxon copleyi*. *s*, septum; *c*, blood capillaries; *a*, air spaces; *arrows*, pneumocytes separating blood capillaries; *m*, smooth muscle; *w*, leukocyte; *e*, red blood cells; *h*, haemocyte. *Scale bar*: **127**, 28 μm; **128**, 6 μm; **129**, 28 μm; **130**, 4 μm; **131**, 7 μm; **132**, 1 μm; **133**, 4 μm

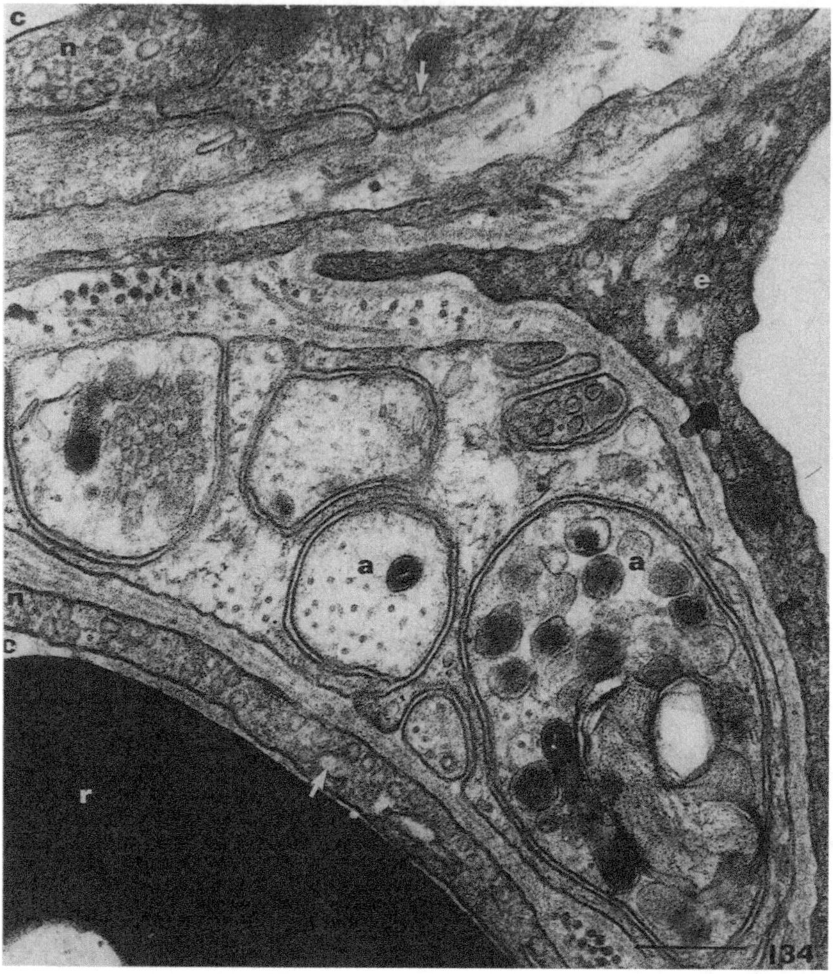

Fig. 134. Lung of a snake, *Dendoraspis polylepis,* showing axonal profiles (*a*) contained in the interstitial space. *c*, blood capillaries; *e*, epithelial cell; *n*, endothelial cell; *arrows*, micropinocytotic vesicles; *r*, red blood cell. *Scale bar*, 1 μm

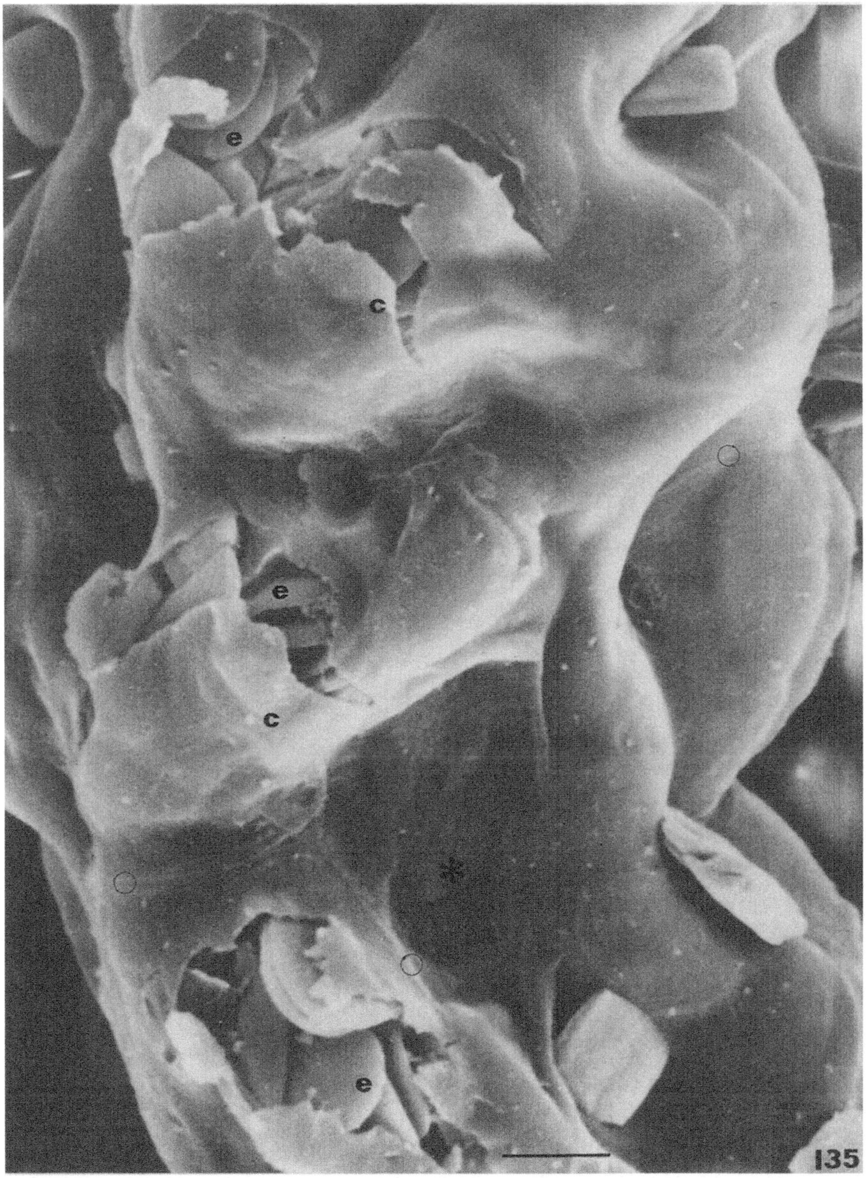

Fig. 135. Surface of an interfaveolar septum of the lung of the snake, *Dendroaspis polylepis*, showing the profuse anastomoses and bulging out of blood capillaries (*c*) into the faveolar air space. *e*, red blood cells; *circles*, junctions of the pneumocytes; *asterisk*, depressions on which the perikarya of the pneumocytes are located. *Scale bar*, 24 μm

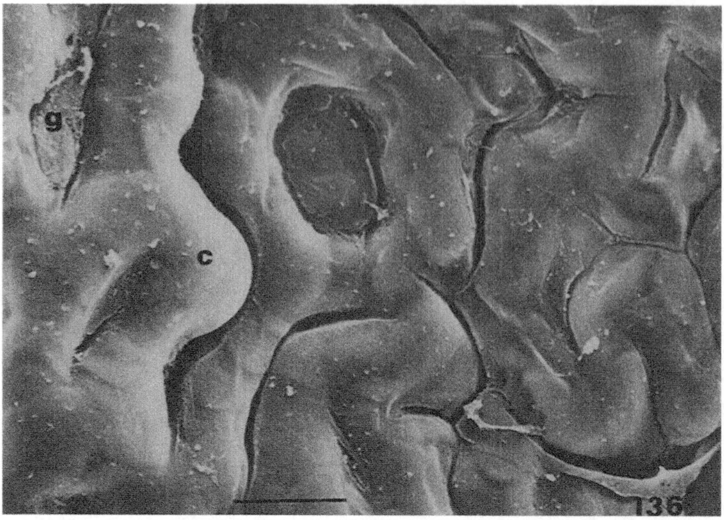

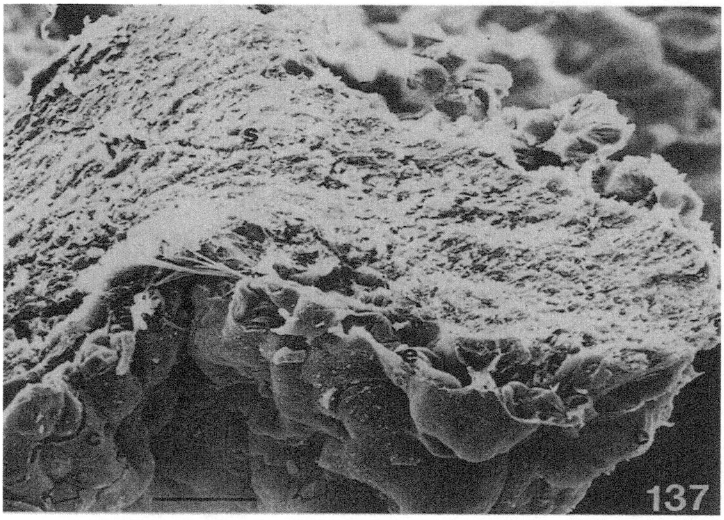

Fig. 136. Anastomoses of blood capillaries (*c*) on the lung of the snake, *Dendroaspis polylepis*. *g*, granular pneumocyte. *Scale bar*, 13 μm

Fig. 137. A septum (*s*) of the lung of the African lungfish, *Protopterus aethiopicus*, showing blood capillaries (*c*) bulging out into the air space. *e*, red blood cell; *arrows*, perikarya of the pneumocytes. *Scale bar*, 37 μm

Figs. 138–140. Tracheobronchial (**138**), arterial (**139**) and venous systems (**140**) of the lung of the domestic pig, *Sus scrofa*. The three systems exhibit a fractal (tree-like) design. *Arrowhead*, tracheal bronchus; *t*, right atrium; *x*, right ventricle. *Scale bars*, 1 cm

Figs. 141, 142. Tracheobronchial (*b*), arterial (*a*) and venous (*v*) systems of the lung of the domestic pig, *Sus scrofa*, showing the fractal nature (tree-like arrangement) and the topological similarities between the three conducting systems. *Arrowhead*, tracheal bronchus; *e*, right ventricle; *t*, right atrium. *Scale bar*, 1 cm

Subject Index

Production: Druckhaus Beltz, Hemsbach

GPSR Compliance

The European Union's (EU) General Product Safety Regulation (GPSR) is a set of rules that requires consumer products to be safe and our obligations to ensure this.

If you have any concerns about our products, you can contact us on ProductSafety@springernature.com

In case Publisher is established outside the EU, the EU authorized representative is:

Springer Nature Customer Service Center GmbH
Europaplatz 3
69115 Heidelberg, Germany

Batch number: 10704713

Printed by Printforce, the Netherlands